Yu Liang

Ultrafast Dynamics of Metalloporphyrins, DNA and Iron-Lanthanide Clusters in the Liquid Phase

Ultrafast Dynamics of Metalloporphyrins, DNA and Iron-Lanthanide Clusters in the Liquid Phase

by
Yu Liang

Dissertation, Karlsruher Institut für Technologie (KIT)
Fakultät für Chemie und Biowissenschaften
Tag der mündlichen Prüfung: 21. Dezember 2012

Impressum

Karlsruher Institut für Technologie (KIT)
KIT Scientific Publishing
Straße am Forum 2
D-76131 Karlsruhe
www.ksp.kit.edu

KIT – Universität des Landes Baden-Württemberg und
nationales Forschungszentrum in der Helmholtz-Gemeinschaft

KIT Scientific Publishing 2013
Print on Demand

ISBN 978-3-86644-999-2

for Ying and Benben

Ultrafast Dynamics of Metalloporphyrins, DNA and Iron-Lanthanide Clusters in the Liquid Phase

Zur Erlangung des akademischen Grades eines

DOKTORS DER NATURWISSENSCHAFTEN

(Dr. rer. nat.)

Fakultät für Chemie und Biowissenschaften

Karlsruher Institut für Technologie (KIT) – Universitätsbereich

genehmigte

DISSERTATION

von

Yu Liang

aus

HAINAN, CHINA

Dekan:	Prof. Dr. M. Bastmeyer
Referent:	PD Dr. A.-N. Unterreiner
Korreferent:	Prof. Dr. R. Schuster
Tag der mündlichen Prüfung:	21.12.2012

Summary

In this work, ultrafast dynamics of different organic and inorganic systems were investigated by means of time-resolved pump-probe spectroscopy in combination with other spectroscopic methods. These dynamics include, for example, the excited state relaxation, rate constants, and mechanisms of charge transfer between donor and acceptor, as well as ultrafast carrier dynamics in nanostructural inorganic clusters. The systems comprise supra-molecular metalloporphyrins, DNA-mediated charge donor and acceptor systems, and inorganic cyclic coordination iron-lanthanide clusters.

Ultrafast Dynamics of Metalloporphyrins

The photophysical properties and ultrafast relaxation dynamics of the first excited state of *meso*-tetra-phenylporphyrinato magnesium (II) (MgTPP) and *meso*-tetra-phenylporphyrinato cadmium (II) (CdTPP) in tetrahydro-furan (THF) have been investigated. In the case of MgTPP, two rapid components on subpicosecond (~100 fs) and picosecond (~3.5 ps) time-scales can be assigned to an ultrafast spectral shift of stimulated emission and vibrational relaxation in the first excited state. The last time constant in the nanosecond regime can be associated to the relaxation to the ground state via internal conversion and fluorescence and/or to higher triplet state via intersystem crossing. The experimental observations suggested that the presence of dark states might influence the depopulation of higher excited states. These states were found at 3.36 eV on the blue side of the Soret-band of MgTPP. For CdTPP, the lifetime of the first excited state (~110 ps) is much shorter than in MgTPP due to fast and efficient intersystem crossing, while the fast component (~5.5 ps) is the time of vibrational relaxation which agrees with that in MgTPP. The absence of the fastest relaxation (subpicosecond timescale) might be caused by the more rigid structure of CdTPP in contrast to MgTPP. From the transient absorption spectra of CdTPP, an additional absorption band at about 3.45 eV revealed an efficient relaxation channel to a higher triplet state, which could be responsible for the short lifetime of the first excited state relative to that of MgTPP proposed by earlier studies. Moreover, a pump-probe response beyond 1600 nm suggests the existence of another relaxation channel besides intersystem crossing with a comparably low lying dark state. TDDFT-calculations show that the dark states originate from an orbital localized on the N-atoms which continuously drops in energy in the series Mg, Zn, Cd. The theoretical results are in good agreement with the experimental findings. In addition, an unusual high transient anisotropy, up to 0.65 in CdTPP, was recorded, which decays on a 100 ps timescale. In contrast to previous works, this behavior was explained by an incoherent mixing of excited state absorption, ground state bleaching, and stimulated emission.

DNA-mediated Charge Transfer

Ultrafast charge transfer dynamics between 6-N,N-dimethylaminopyren-modified 2'-deoxyuridine (**Ap-dU**) as charge acceptor and Nile red-modified 2'-deoxyuridine (**Nr-dU**) as charge donor through a DNA double helix has been studied. The efficiency of the fluorescence quenching in **Nr-dU** depends on two components: a) the presence of the charge acceptor **Ap-dU**; and b) the number of intervening base pairs between the donor and the acceptor. This observation clearly indicates a certain distance-dependence of the photoinduced charge transfer between **Nr-dU** and **Ap-dU**. However, the quenching efficiency is not linearly, or exponentially, dependent on the distance between donor and acceptor (e.g., 73% for two intervening C:G base pairs and the same for two intervening A:T base pairs, 68% for one intervening C:G base pair and 37% for one intervening A:T base pair, less than 10% for three base pairs). A comparison with the results of time-resolved measurements revealed a correlation between quenching efficiency and the first ultrafast time constant, suggesting that quenching proceeds via a charge transfer from the donor to the acceptor. The experimental results gave two decays with a rapid charge transfer rate of about 600 fs ($\sim 10^{12}$ s^{-1}) that depends strongly, and nonlinearly, on the distance between donor and acceptor, and a slower time constant of a few picoseconds ($\sim 10^{11}$ s^{-1}) with weak distance-dependence. However, both time constants are slightly temperature-dependent. The third time constant on a nanosecond timescale represents the fluorescence lifetime of the charge donor **Nr-dU**. According to the experimental findings and theoretical calculations (TDDFT), a combination of single-step super-exchange and multistep hopping mechanisms can be proposed for this short range charge transfer system. Furthermore, significantly less quenching efficiency and slower charge transfer rates at very short distances indicate that the direct interaction between donor and acceptor molecules leads to a local structural distortion of DNA duplexes, which may provide some uncertainty in identifying the charge transfer rates in this system.

Ultrafast Carrier Dynamics in Iron-Lanthanide Clusters

The ultrafast carrier dynamics in an inorganic nanostructural cyclic coordination cluster system $[Fe^{(III)}_{10}Ln^{(III)}_{10}(Me\text{-}tea)_{10}(Me\text{-}teaH)_{10}(NO_3)_{10}] \cdot n\text{MeCN}$ (**Fe$_{10}$Ln$_{10}$**, Ln = Y, Pr, Nd, Sm, Eu, Gd, Tb, Dy, Ho, Er, Tm, Yb, Lu) has been investigated. Generally, for all **Fe$_{10}$Ln$_{10}$** clusters, three ultrafast time constants measured can be assigned to different processes: photoexcited electrons relax to the edge of the conduction band as well as into trap states caused by doped lanthanide ions with a time constant of about 200-400 fs. Electrons subsequently leave the conduction band and primarily decay via nonradiative recombination within a few picoseconds. The lifetime of the trap states is on the order of 100 ps. The experimental results were found for the excited states close to that observed with $[Fe_6(tea)_6]\cdot 6MeOH$ (**Fe₆**-cluster) taken as reference compounds and are in good agreement with that of Fe_2O_3 nanoparticles studied previously. Based upon the experimental results and literature suggestion, the ultrafast relaxation processes can be proposed to the ligand to metal charge transition (LMCT) inside the Fe-O domain, while the doped lanthanide ions inside the cluster influence this dynamics. The influence of doped lanthanide ions on the decay time and corresponding relative amplitudes exhibits a complicated behavior. However, the good correlation between the first relative amplitudes and the optical energy gaps indicates that the larger the energy gap is, the more efficient hot electrons trap. Further photolysis experiments revealed the presence of bivalent iron ions (Fe^{2+}) in the cluster after irradiation. This observation is based on the photoinduced reaction between **Fe₁₀Ln₁₀** clusters and potassium ferricyanide. After irradiation, a deep blue color precipitation was obtained in solution within 1 second. A comparison with the results of steady-state absorption and time-resolved measurements between Prussian blue and the deep blue color production revealed the presence of a long lifetime metal to metal charge transfer (MMCT) state, indicating that the irradiated compound contained a Fe^{3+}-CN-Fe^{2+} unit with spectral properties similar to Prussian blue analogous.

Contents

Summary..i

List of Abbreviations ix

Chapter 1: General Introduction 1

Chapter 2: Ultrafast Laser Spectroscopy 5

    2.1 Theoretical Aspects 6

        2.1.1 Nonlinear Optics..6

        2.1.2 Second-order Susceptibility 8

    2.2 Femtosecond Laser System 11

        2.2.1 The Ti:sapphire Oscillator 12

        2.2.2 Regenerative Amplifier (RGA) 15

        2.2.3 Non-collinear Optical Parametric Amplifier (NOPA).....17

    2.3 Pump-probe Spectroscopy..18

Chapter 3: Ultrafast Dynamics and Transient Anisotropy of Metalloporphyrins 21

    3.1 Introduction..22

    3.2 Experimental Setup and Conditions 24

    3.3 Steady-state Spectroscopic Properties of Porphyrins 26

        3.3.1 UV/Vis Absorption Spectroscopy 26

        3.3.2 Solvent Effects 30

        3.3.3 Fluorescence Emission Spectroscopy 34

        3.3.4 Fluorescence Excitation Spectroscopy....................37

    3.4 Ultrafast Relaxation Dynamics of the First Excited State.........41

3.4.1   Relaxation Dynamics of MgTPP 42

3.4.2   Relaxation Dynamics of CdTPP 44

3.4.3   The Role of Dark States 46

3.5   Transient Anisotropy..55

3.5.1   Anisotropy for Nondegenerate and Degenerate States....56

3.5.2   Anisotropy for CdTPP in THF 59

3.5.3   Simulation of the Transient Anisotropy 61

3.6   Conclusion and Outlook 64

Chapter 4: Ultrafast Laser Spectroscopy 65

4.1   Introduction 66

4.2   Charge Transfer Mechanisms and Rates 69

4.2.1   Single-step Superexchange 70

4.2.2   Multistep Hopping 71

4.2.3   Boundary of Superexchange and Hopping Model...........72

4.3   Spectral Properties of the Donor-DNA-acceptor System 74

4.4   Charge Transfer through DNA 79

4.5   Conclusion and Outlook 92

Chapter 5: Ultrafast Carrier Dynamics in Iron-lanthanide

Clusters 93

5.1   Introduction 94

5.2   Experimental Setup 95

5.3   Structure of Iron-lanthanide Clusters.......................................96

5.4   Reference Systems 98

5.5   UV/Vis Absorption Spectroscopy 99

5.6   Ultrafast Dynamics in Iron-lanthanide Clusters 102

5.7   Energy Gaps 110

5.8   Photolysis of Iron-lanthanide Clusters 114

5.8.1 Irradiation Experiments 115

5.8.2 UV/Vis Absorption Spectroscopy of Irradiation Product
....117

5.8.3 Pump-probe Spectroscopy of the Irradiation Product....119

5.9 Conclusion and Outlook...121

Appendix **123**

A.1 The femtosecond laser system...124
A.2 The associated system for the femtosecond laser system 125
A.3 Calibration of spectral peaks in Vis- and NIR-regions...........126
B.1 Spectral narrowing of the probe pulse...127
B.2 UV/Vis absorption spectra of CdTPP in THF after different
irradiation times...127

B.3 Transient response of CdTPP in THF in a standing cuvette 128
C.1 Probe wavelengths dependence of $\mathbf{Fe_{10}Tm_{10}}$ clusters after
excitation at 310 nm 128

C.2 UV/Vis absorption spectra of $\mathbf{Fe_{10}Er_{10}}$ clusters after different
irradiation times...129

C.3 Probe wavelengthsdependence of $\mathbf{Fe_{10}Tm_{10}}$ clusters after excitation at 310 nm 129

Bibliography...**131**

Acknowledgments **157**

List of publications **159**

List of Abbreviations

Ap	6-N,N-dimethylaminopyrene
A:T	adenine thymine base pair
B3LYP	hybrid exchange-correlation functional consisting of B88 and LYP
BBO	β-barium borate
BP86	exchange-correlation functional consisting of B88 VMN and P86
bps	base pairs
C:G	cytosine guanine base pair
CPA	chirped pulse amplification
CT	charge transfer
CW	continuous wave
ΔOD	the change of optical density
DFG	different-frequency generation
DFT	density functional theory
DNA	deoxyribonucleic acid
dU	deoxyuridine
EDA	energy decomposition analysis
ESA	excited state absorption
FWHM	full width at half maximum
FWM	four-wave mixing
GSB	ground state bleaching
HOMO	highest occupied molecular orbital

IC	internal conversion
ISC	intersystem crossing
KLM	Kerr lens mode locking
KTP	potassium titanyl phosphate
λ_p	the central wavelength of the pump pulse
λ_{pr}	the central wavelength of the probe pulse
LBO	lithium borate
LMCT	ligand to metal charge transfer
LUMO	lowest unoccupied molecular orbital
MMCT	metal to metal charge transfer
MO	molecular orbital
Nd:YAG	neodym doped yttrium aluminum garnet
NOPA	non-collinear optical parametric amplifier
NIR	near infrared
Nr	Nile red
OD	optical density
OPA	optical parametric amplification
OR	optical rectification
PD	photodiode
RGA	regenerative amplifier
SE	stimulated emission
SHG	second harmonic generation
SFG	sum-frequency generation
SMM	single molecule magnet
TDDFT	time-dependent density functional theory
THF	tetrahydrofuran
THG	third harmonic generation

Ti:sapphire	titanium doped sapphire
TPP	tetraphenylporphyrin
UV	ultraviolet light
Vis	visible light
VR	vibrational relaxation
Vs. NHE	versus normal hydrogen electrode

Chapter 1: General Introduction

The general definition of spectroscopy is the study of the interactions between light and matter and provides a wealth of knowledge, for example, about the information of electronic structure in atoms, molecules, semiconductors, etc. However, most photophysical and photochemical processes in chemistry are chemical bonds breaking and changing in reaction, internal electronic motion, and electron transfer between two (redox) partners. Such physical events happen usually on the picosecond or even on the femtosecond time scale (1 fs = 10^{-15} s). Hence, a femtosecond time resolution of the experiment is required to observe those motions directly in real time. Ultrafast spectroscopy involves temporally short light pulses, which can be employed to gain insight into the ultrafast dynamics on femtosecond time scale. The femtosecond laser spectroscopy has opened up a range of applications for many branches of physics, chemistry as well as biology.[1-3] Thus, in chapter 2, an overview of the fundamental principles of femtosecond laser systems used for time-resolved pump-probe experiments will be briefly represented. Using this femtosecond laser system several organic, biological, and inorganic systems will be investigated to observe the ultrafast phenomena.

The first investigated molecule is metalloporphyrin. Metalloporphyrins are highly stable macrocyclic π-systems that display interesting photophysical and photochemical properties. These properties make them an essential constituent for biological processes. For instance, chlorophyll with magnesium as a central atom are well known for the conversion of sunlight into energy in plants, whereas hemoglobin (iron as central atom) transports oxygen in blood.[4] Also, metalloporphyrins have potential applications in artificial solar energy harvesting systems and molecular photonic materials.[5-7] These processes and applications are generally related to their high stability and efficient light absorption ability in the visible and near infrared region of the optical spectrum. Therefore, to acquire a deeper knowledge of these ultrafast processes, as well as related photophysical properties, in chapter 3, the ultrafast relaxation dynamics on excited states of monomer metalloporphyrins has been investigated in detail using time-resolved pump-probe techniques. Typically, the ultrafast relaxation dynamics of a photoexcited molecular system can be described based upon the Jabłoński diagram.[8,9] Once a molecule has absorbed photon energy, there are a number of pathways to return to the ground state (S_0). If the electronic state is not changed, but there is a coupling to some manifold of vibrational states, intramolecular vibrational relaxation (VR) can be observed.[10] If there is a coupling to another electronic state with the same spin state (e.g. S_1-S_0), nonradiative intramolecular internal conversion (IC) or radiative fluorescence (FL) takes place. If the spin states of the initial energy levels are changed, the molecule undergoes an intersystem crossing (ISC) from a singlet state to a triplet state, and then phosphorescence can occur radiatively via triplet-singlet transitions. These photophysical processes are radiative or radiationless transitions inside the molecule but do not lead to structural changes, although the bond lengths and angles differ in different excited electronic states.[9] The ultrafast relaxation dynamics on the first (Q-band) and the second excited (Soret-band) states of many metalloporphyrin derivatives such as ZnTPP, MgTPP and CdTPP have been studied for a long time.[11,12] However, little studies were focused on the dark states which can influence the relaxation dynamics.[13] Moreover, evidence for the existence of dark states has been given theoretically and experimentally. Based upon the experimental findings and calculations, a model for the relaxation pathway of excited states is presented in chapter 3.

Photoinduced charge transfer (CT) plays an important role in the physical, chemical and biological sciences. Those transfer processes in the living plant, or in the human body, provide many insights into the chemical and biological reactions. The charge transfer in double stranded deoxyribonucleic acid (DNA) is one of the most common topics. DNA has a well-defined π-stack structure. This structure refers to attractive, non-covalent interactions between the base pairs. This structural property leads to considerable interest in charge transport through DNA and the molecule's potential use as molecular wire in nanoelectronics.[14,15] The studies on the photoinduced, or photoinitiated, charge transfer processes in DNA were generated during the last two decades.[16,17] The goal of those studies was to better understand oxidative damage within the cell and potential applications in the development of DNA-based molecular wires.[18-20] For the oxidative type, the photoinduced electron hole transfer via two different mechanisms were elucidated theoretically[21] and experimentally[22]. In general, for short range (intervening base pairs n $\leq$ 3) the transfer occurs coherently via superexchange mechanism, while long range hole transfer can only be explained by an incoherent hopping model. However, from a theoretical point of view, charge hopping could occur potentially also over very short distances in DNA. Hence, in chapter 4, a new synthetic donor-DNA-acceptor system was employed to study the charge transfer rates and mechanisms over short range (n $\leq$ 3) using time-resolved pump-probe technique.

Another group of chemical systems that has been chosen for investigation are heterometallic iron-lanthanide coordination compounds. These compounds have some interesting magnetic properties such as single molecule magnet (SMM) behavior,[23,24] which can be proposed as potential nanomagnets for high-density information storage and processing.[25,26] Many studies focused on SMMs based on the combination of $3d$ and $4f$ ions.[25-29] In these heterometallic solids, f-f interactions are weak due to shielding effects of outer s- and p-electrons. However, interaction of f-electrons with the more expanded s-, p-, or d-electrons should be relatively stronger.[30] Trivalent Fe ions as transition metal with $3d$-shell electrons are usually used in iron-type SMMs due to their high magnetic anisotropy and high spin quantum numbers. Meanwhile, the advantage of lanthanide ion's significant spin and/or its large anisotropy in a mixed $3d/4f$ metal complexes is a useful way to generate SMMs with properties distinctly different from those of the homometallic ones.[31,32] Indeed, families of Fe^{3+}- and Ln^{3+}-containing compounds have been synthesized and studied in order to assess various contributions to magnetic properties, resulting from differing Ln^{3+} electronic configurations and geometrical changes.[33,34] Bulk magnetic susceptibility and Mössbauer spectroscopic studies showed that the low temperature magnetic behavior can arise from the interaction of strongly antiferromagnetically coupled iron subchains with the lanthanide ions, and remarkably differs in the lanthanide series. However, from a photophysical point of view, the exchange interactions might be cause by a shift of electron density from the $3d$ orbitals of transition metal ions to empty $5d$ orbitals of lanthanide ions.[35-37] Therefore, pump-probe absorption measurements are desirable to get more insight into such ultrafast electron transfer. The results will be discussed in chapter 5.

Chapter 2: Ultrafast Laser Spectroscopy

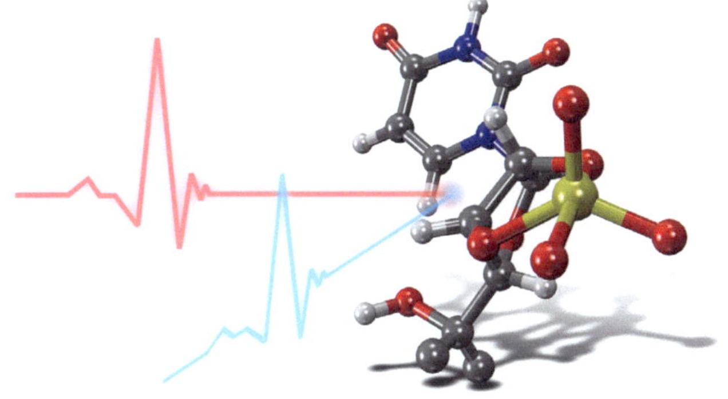

2.1 Theoretical Aspects

The advent of femtosecond laser systems and the extension of laser wavelengths by frequency conversion techniques provide a variety of extremely short light pulses to investigate ultrafast phenomena in physical, chemical, and biological systems extensively and successfully. The knowledge of nonlinear optics is important and necessary not only for understanding femtosecond laser systems, but also for the development of short-pulse lasers. Laser light with sufficient intensity enables one to examine the modification of the optical properties of a material system. In this section, a brief introduction to nonlinear optical phenomena is presented.

2.1.1 Nonlinear Optics

In classical physics, the interaction between light and matter is described by Maxwell's equations. The solutions of these equations can predict the interactions of electromagnetic fields with arbitrary media such as propagation, attenuation, emission, amplification, as well as scattering.[38] In the case of linear optics, the induced polarization ($\vec{P}(t)$) which increases proportionally as a function of the electric filed strength ($\vec{E}(t)$), can be described by a linear function:

$$\vec{P}(t) = \epsilon_0 \chi^{(1)} \vec{E}(t) \tag{2.1}$$

where ϵ_0 is the permittivity in vacuum and $\chi^{(1)}$ is the linear susceptibility of the medium. In the domain of linear optics, the optical properties of a medium are independent of the electric field strength. As a consequence, the refractive index in this case is clearly independent of the applied field. The frequencies of the incident light remain constant. In nonlinear optics, on the other hand, when the light is sufficiently intense compared to the intensity of electric field in the internal atoms or molecules (e.g., laser beam), the polarization may respond nonlinearly to the field strength and generate higher harmonics. The response can be described as a Taylor expansion of the material polarization in power series of the field strength[39]

$$\vec{P}_i(t) = \epsilon_0\left[\chi_{ij}^{(1)}\vec{E}_j(t) + \chi_{ijk}^{(2)}\vec{E}_j(t)\vec{E}_k(t) + \chi_{ijkl}^{(3)}\vec{E}_j(t)\vec{E}_k(t)\vec{E}_l(t) + \cdots\right]$$

$$= \vec{P}_i^{(1)}(t) + \vec{P}_i^{(2)}(t) + \vec{P}_i^{(3)}(t) + \cdots \tag{2.2}$$

$$= \vec{P}_i^{L}(t) + \vec{P}_i^{NL}(t).$$

The coefficients $\chi^{(n)}$ are known as the optical susceptibilities of the n-th order nonlinear process. The first term $(\vec{P}_i^{L}(t) = \epsilon_0\chi_{ij}^{(1)}\vec{E}_j(t))$ represents the linear optical processes. The residual terms of the polarization $\vec{P}_i^{NL}(t)$ in Eq. (2.2) exhibit the nonlinear response of the polarization $\vec{P}(t)$ with respect to the field strength $\vec{E}(t)$. Nonlinear effects result in a series of the order of susceptibilities $\chi^{(n)}$, which are based on the nonlinear modulation of the dielectric constant of the material by the incident light as a result of the strong forces from the interaction between the electric light field vector and the electrons of the matter.

One of the most common nonlinear optical processes with second order nonlinear susceptibility $\chi^{(2)}$ is the second harmonic generation (SHG) of the incident light. The first experimental evidence of SHG was obtained by Franken et al.[40,41] They focused a ruby laser beam with wavelength 694.3 nm on a quartz crystal and observed one of the two outgoing beams in the UV region (wavelength at 347.1 nm) from the crystal. This report gave rise to the discovery of a rich diversity of nonlinear optical effects, including sum and difference frequency generation (SFG and DFG) and optical rectification (OR)[42]. In the case of higher nonlinear susceptibility (e.g.,

$\chi^{(3)}$), the well-known nonlinear processes are self-focusing, four-wave mixing (FWM), and so on. The most prominent examples of nonlinear effects include Pockels and Kerr electrooptics.[43]

2.1.2   Second-order Susceptibility

Generally, the electric field strength $\vec{E}(t)$ of a high intensity incident light is presented by angular frequency ω and complex amplitude A

$$E(t) = Ae^{-i\omega t} \ . \tag{2.3}$$

Based on the second-order susceptibility $\chi^{(2)}$, the nonlinear polarization generated in this medium can be described as follows according to the equation (2.2) together with (2.3):

$$P^{(2)}(t) = \epsilon_0 \chi^{(2)} E^2(t)$$

$$= 2\epsilon_0 \chi^{(2)} A^2 + \epsilon_0 \chi^{(2)} A^2 e^{-i2\omega t} \ . \tag{2.4}$$

Obviously, the second-order polarization consists of two contributions: a first term with zero frequency and a second term at frequency 2ω. The latter contribution leads to the generation of frequency doubling (2ω). The first term corresponds to a steady component and leads to another process known as optical rectification which can create a static potential difference across the nonlinear medium. An important symmetry aspect of the Taylor expansion by (2.2) is that all even-order coefficients must vanish for media with inversion symmetry. The SHG-process therefore can take place only in media with no inversion symmetry. Note that most materials do have inversion symmetry and SHG-processes can only be observed in a very specific class of nonlinear crystals.

One of the conditions to increase the conversion efficiency of the SHG-process is known as the phase-matching condition in nonlinear optics.[44] This condition represents momentum conservation for the SHG-process by generating a single photon of the second harmonic with two photons of the incident field. Since the wave vectors $\vec{k}_1$ and $\vec{k}_2$ are corresponding to the momenta of the incident and the second harmonic fields

$$\vec{p}_1 = \frac{h}{2\pi}\vec{k}_1 \qquad \text{and} \qquad \vec{p}_2 = \frac{h}{2\pi}\vec{k}_2 \tag{2.5}$$

with h the Planck constant. Based upon momenta conservation ($\Delta\vec{p} = \vec{p}_1 + \vec{p}_2 = 0$) one can obtain the condition of perfect phase matching ($\Delta\vec{k} = 0$). The most common approach for achieving phase matching is to use birefringent nonlinear (also anisotropic) crystals.[45,46] All isotropic crystals have equivalent axes that interact with light in a similar manner, regardless of the crystal orientation with respect to the incident light.[47] Anisotropic crystals, on the other hand, have crystallographically distinct axes and interact with the incident light depending upon the orientation of the crystalline lattice with respect to the incident light angle.[47] When light travels through a birefringent crystal, the waves are split into two components (so-called ordinary- and extraordinary wave). Ordinary waves are polarized perpendicular to the optical axis while extraordinary waves are polarized parallel to the optical axis. Figure 2.1 illustrates the phase matching in a birefringent crystal. The circle exhibits the cross section of the refractive index for an ordinary wave (n_o) with frequency ω. The ellipse represents the refractive index cross section for the extraordinary wave (n_e) with the frequency of 2ω. Phase matching is achieved when the condition $n_o(\omega) = n_e(2\omega)$ is fulfilled.[48] The angle θ is the angle between the direction of propagation and the optical axis.

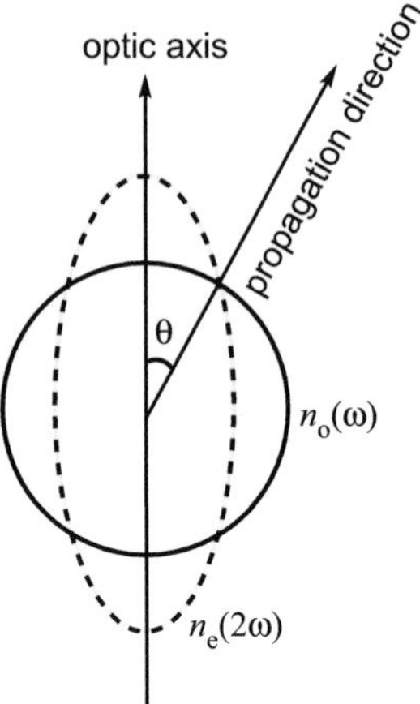

Figure 2.1: Phase matching in a birefringent crystal.[48]

The second-order nonlinear polarization defined by (2.4) gives rise to three-wave mixing processes and optical rectification. In particular, when the incident light consists of two distinct frequency components ω_1 and ω_2, which can be represented in the form

$$E(t) = A_1 e^{-i\omega_1 t} + A_2 e^{-i\omega_2 t} \tag{2.6}$$

using the second-order contribution to the nonlinear polarization one can obtain the solution as

$$P^{(2)}(t) = \epsilon_0 \chi^{(2)} E^2(t)$$

$$= \epsilon_0 \chi^{(2)} (A_1^2 e^{-2i\omega_1 t} + A_2^2 e^{-2i\omega_2 t} + 2A_1 A_2 e^{-i(\omega_1+\omega_2)t}$$

$$+ 2A_1 A_2 e^{-i(\omega_1-\omega_2)t}) + 2\epsilon_0 \chi^{(2)} (A_1 A_1 + A_2 A_2). \tag{2.7}$$

Eq. (2.7) shows the various frequency components with different amplitudes. Those contributions of nonlinear polarization consist of various nonlinear processes, such as second-harmonic generation (SHG, $2\omega_1$ and $2\omega_2$), sum-frequency generation (SFG, $\omega_1 + \omega_2$), difference-frequency generation (DFG, $\omega_1 - \omega_2$), as well as optical rectification (OR, the last term). In many ways the process of SFG is analogous to that of SHG, except that in SFG the two input waves are at different frequencies. One application of SFG is to produce tunable radiation in the ultraviolet spectral region by mixing of two input pulses at different angular frequencies ω_1 and ω_2. One of the input waves is the output of a fixed-frequency visible laser pulse, while the other is the output of a frequency-tunable visible laser pulse (see experimental section in chapter 5). DFG, on the other hand, can be used to generate tunable infrared pulses by mixing the output of a frequency-tunable visible pulse with that of another fixed-frequency visible pulse. Again the efficiency of these processes depends on the phase matching condition.

Some nonlinear materials such as BBO (Barium borate, $\beta\text{-BaB}_2\text{O}_4$), LBO (Lithium borate, LiB_3O_5), and KTP (KTiOPO_4) are applied in tunable laser frequency conversion, for example, SHG.[49] These crystals possess high nonlinearity, high resistance to laser damage, and large birefringence. BBO is one of the most common in use nonlinear crystals in Nd:YAG laser for harmonic generation and optical parametric amplification (OPA). The conversion efficiency is up to 70% for SHG, 60% for THG.[49]

2.2 Femtosecond Laser System

With the help of nonlinear processes, the width of laser pulses can be brought down from picoseconds to femtoseconds. Nowadays femtosecond pulses in the range of 10 fs, and even below, can be generated from reliable laser oscillators. Such femtosecond laser allows photochemistry processes and molecular dynamics to be monitored in real time. A femtosecond laser system in the institute of physical chemistry in university Karlsruhe (TH) was built by Priv.-Doz. Dr. A.-N. Unterreiner and described exhaustively in his dissertation.[50] After being moved into another building, the system had to be rebuilt and mounted on a new optical table (Newport). Thus, in this section, three basic building blocks of this laser system are briefly summarized. The operating principles of Ti:sapphire oscillators are described in the first part of this section. The second part gives an introduction into the amplification of laser pulses with regenerative amplification techniques. The last part shows how to generate a tunable laser pulse using non-collinear optical parametric amplifier (NOPA). The whole femtosecond laser system is illustrated in Appendix A.1 and the related system is drafted in Appendix A.2.

2.2.1 The Ti:sapphire Oscillator

Titanium-doped sapphire (Ti:sapphire) is a solid state laser medium capable of tunable laser operation over a broad range of visible and near infrared (NIR) wavelengths. Because of its broad absorption band in the blue and green, the energy for lasing action can be supplied by standard continuous wave (CW) argon ion lasers with output wavelength of 532 nm (Figure 2.2). Based upon nonlinear processes, when a laser beam propagates in a nonlinear material, the refractive index which depends nonlinearly on the propagating field and can be described by[51]:

$$n = n_0 + \frac{1}{2}n_2 I \qquad (2.8)$$

where n_0 is the refractive index at low intensities and n_2 the nonlinear index coefficient describing the coupling between the electric field and the refractive index.[52] I is the intensity of the laser beam. Thus, the refractive index is as a function of intensities along the optical pathway. Since the nonlinear index coefficient is usually positive for most materials, the refractive index becomes higher at the centre of the beam than at the edges. This beam creates a gradient index (GRIN) lens in the material and leads to self-focusing of the beam.[53] In the laser cavity, self-focusing in a gain medium can lead to an effective fast saturable absorber that can be used to suppress CW operation into a short intense pulse (Figure 2.3). This process is known as the Kerr lens effect.[54]

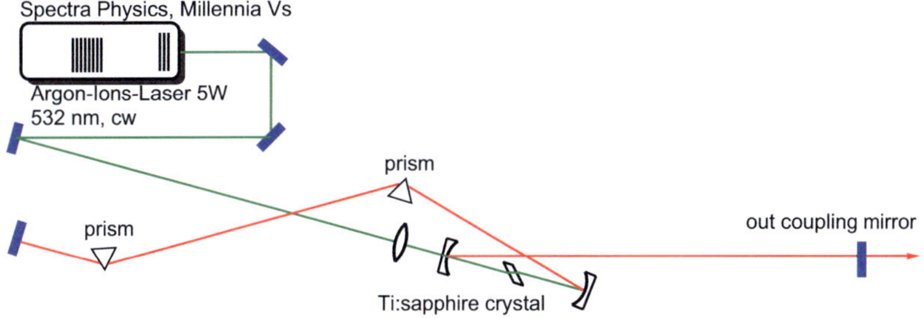

Figure 2.2: Schematic diagram of a Kerr lens model-locked Ti:sapphire laser.

In a laser, many modes are separated in frequency by

$$\Delta \nu = \frac{c}{2L} \tag{2.9}$$

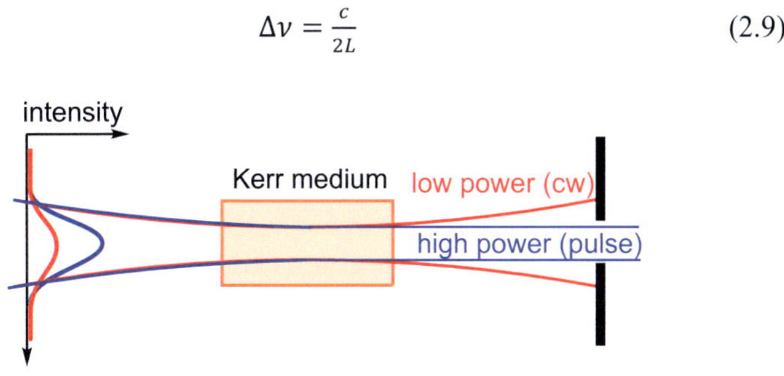

Figure 2.3: Illustration of a Kerr lens mode-locked Ti:sapphire laser.[53]

where L is the length of the laser cavity and c is the speed of light. These modes usually oscillate with random phases with irregular amplitudes. These result in randomly time-varying amplitude within the round-trip period

$$T = \frac{1}{\Delta v} = \frac{2L}{c}.$$

(2.10)

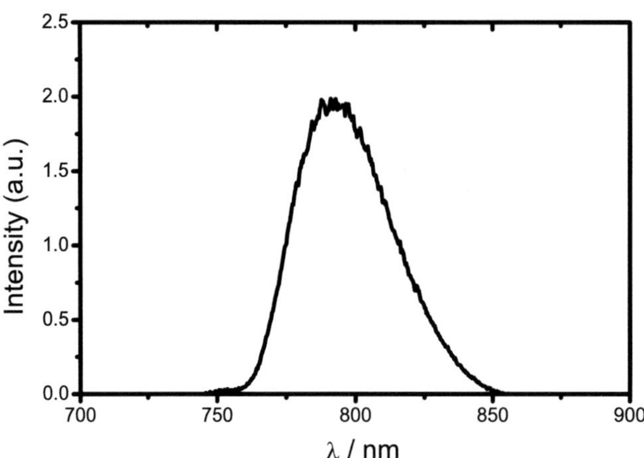

Figure 2.4: Spectrum of mode-locked Ti:sapphire oscillator with a central wavelength at 800 nm detected using spectrometer (Ocean Optics).

However, if these modes oscillate with the same initial phase (phase matching), and the time distribution shows a periodic repetition of a wave packet, they will constructively interfere to each other and cause the beam to produce a train of pulses. The laser is said to be phase-locked or mode-locked. Thus, the width of the pulses is mainly determined by the gain medium of the laser and the length of the cavity. A prim pair is applied to compensate the material dispersion of one round-trip in the cavity. Based upon the mode-locking technique, a modulation is generated via moving a prism in or out of the resonator. This modulation can be automatically synchronized to the cavity round-trip frequency. The length (L) of the resonator dominates the repetition rate. In this laser system, the Ti:sapphire oscillator operates at a repetition rate of 93 MHz and provides high output energy at about 420 mW with a pump energy of 3.80 W. The duration of the output pulses is about 25 fs with a central wavelength at 800 nm (Figure 2.4).

2.2.2 Regenerative Amplifier (RGA)

Regenerative amplification (RGA) of femtosecond laser pulses also takes place in a Ti:sapphire crystal in a resonator pumped by a Nd:YAG laser with a repetition rate of 1 KHz (see Figure 2.5). A detail description of the regenerative amplification can be found in ref. [50] and [55]. Before the seed pulses from Ti:sapphire oscillator enter the resonator, the intensity of the high energy pulses need to be lowered via an achromatic, diffraction-grating pulse stretcher in order to avoid damage of the optics in the RGA system caused by highly intense pulses. The principle of the pulse stretcher was described in Ref. [50]. The principle of RGA is based on the trapping of a pulse in the laser resonator. After trapping, the pulse is kept until its energy increases in the amplification cavity. Here a Pockels cell (Medox) in combination with a thin film polarizer (TFP, Newport) acts as an optical switch allowing the in- and out-coupling of the pulses. The latter optics (TFP) can reflect s-polarized and transmit p-polarized pulses. The Pockels cell consists of a nonlinear optical crystal with the voltage applied to the crystal. The crystal becomes birefringent under the influence of the applied electric field. The Pockels cell can be set in two stages, corresponding to $\lambda/4$-plate without applying of electric field and non-birefringent material with the influence of the voltage.[56] When a s-polarized pulse (perpendicular to the plane of the optical table) enters the resonator. This s-polarized pulse will be reflected via the thin film polarizer and passes through the

Pockels cell (initially working like a $\lambda/4$-plate). The pulse passes the Pockels cell twice via a reflective mirror, changing the polarization of the pulse to p-polarization (parallel to the plane of optical table). The p-polarized pulse can transmit through TFP and enter the cavity. At the same time this pulse triggers the applied electric field. The Pockels cell is switched on and acts as a non-birefringent material, which means that the Pockels cell has no effect on the polarization of the pulse. During the period that the p-polarized pulse travels through the cavity, the pulse is kept in the cavity and amplified. After the intensity of the pulses reaches the desired level, the voltage on the Pockels cell is switched off again and makes the Pockels cell as $\lambda/4$-plate. The pulse passes the Pockels cell twice and changes the polarization from p to s. Now the TFP can reflect the s-polarized pulse, coupling the pulse out of the cavity. A Faraday rotator is used for separating input and output pulses. After the amplification stage the pulses are re-compressed to a duration of about 70 fs (Figure 2.6). The energy content of the amplified pulses is about 450 μJ and the spectral width 30 nm (FWHM).

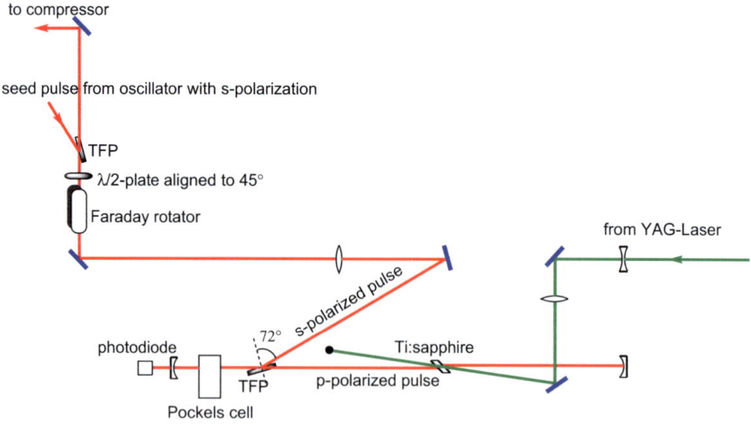

Figure 2.5: Regenerative amplifier of a femtosecond laser system.

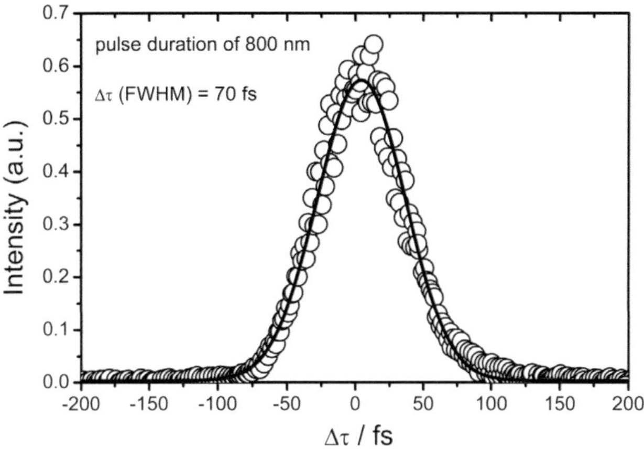

Figure 2.6: Characterization of the pulse duration ($\Delta\tau$ (FWHM) = 70 fs) at a center wavelength of 800 nm using intensity autocorrelation.

2.2.3　Non-collinear Optical Parametric Amplifier (NOPA)

To access various spectral regions, nonlinear optical processes are used. In addition to second harmonic generation, the non-collinear optical parametric amplification (NOPA) in a nonlinear crystal (here BBO, Döhrer Elektrooptik) is one of the most common tools used to generate ultrafast pulses at various wavelengths.[57-60] Figure 2.7 shows a scheme of a home-built NOPA system, which consists of three components: white light generator, first- and second-stage for amplification. The white light generator, namely white light continuum, is used as a tunable ultrafast light source. A small fraction (2%) of the output pulse from the femtosecond laser is focused into transparent materials (here a sapphire plate, 2 mm). As a result of nonlinear optical processes, the short pulse produces a broadened spectrum. Most of the energy is unchanged around the fundamental frequency of the femtosecond laser, but a broad intensity distribution of the light covers the wavelengths from 450 to 900 nm. Two subsequent stages for parametric amplification of the seed white light in BBO crystals are pumped by the doubled frequency of the fundamental output (the 800 nm beam). The output signal pulses are therefore tunable throughout the visible and the near infrared (470-1600 nm). The amplification provides pulse durations of about 100 fs after compression with a prism pair, and 2 μJ pulse energy. To calibrate the pulse wavelength a series of interference filters (ThorLabs) is applied and the calibration is shown in Appendix A.3.

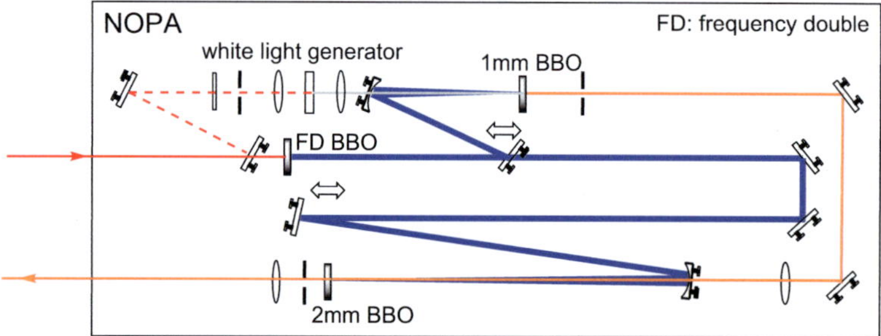

Figure 2.7: Non-collinear optical parametric amplifier (NOPA).

2.3 Pump-probe Spectroscopy

The principle of the pump-probe technique is that one pulse (~1 μJ) is used to excite a system of interest and trigger the ultrafast processes under investigation, while a second pulse with a relative weak energy (< 0.1 μJ), which arrives at a variable delay time, is used to probe how the system has been altered by the pump pulse. The probe beam was detected by silicon photodiodes in the visible and by InGaAs photodiodes in the near infrared region. Using two identical detectors, the intensities of the probe beam in front of (I_0) and behind (I) the sample were recorded to give the relative change of the optical density (ΔOD) with and without pump pulse:

$$\Delta OD = \log\left(\tfrac{I_0}{I}\right)_w - \log\left(\tfrac{I_0}{I}\right)_{w/o}. \tag{2.11}$$

By changing the delay between two pulse trains, one can obtain a spectral response of the sample's excited state absorption (ESA), stimulated emission (SE), ground state bleaching (GSB), or combination of each other. The ESA shows positive absorption responses ($\Delta OD > 0$) corresponding to resonant transitions of the states that are populated after optical excitation via absorption of the probe photons by the molecule (Figure 2.8a). A negative response ($\Delta OD < 0$) can be assigned to SE or GSB where the intensity of the transmitted probe light increases (Figure 2.8b and 2.8c). The evolution of the induced bleaching/absorption amplitude with the delay time provides information about the lifetime of the excited state. In pump-probe experiments the time resolution is usually limited by the pulse duration. The pump-probe experimental setup is schematically presented in Ref. [61] and also in Appendix A.1. In this femtosecond laser system, the probe pulses were delayed with respect to the pump pulses using a computer controlled transition stage (Melles Griot). In order to ensure that the probe pulse was completely located within the excitation range of the pump pulse in the sample, its diameter (0.5 mm for probe and 1.5 mm for pump pulses) was smaller than half of the pump beam. Pump and probe beams intersected at a small angle of about $4°$ in the sample. In order to minimize anisotropic effects, the polarization of the probe pulse was set at the magic angle with respect to the pump pulse using a tunable $\lambda/2$-plate (Alphalas).

Two typical transient response curves are shown Figure 2.8 (at the bottom). The peaks around zero time delay come from the overlap of the pump and probe pulses in the sample. With the help of an error function an exponential decay function, which is used to obtain the decay time constants, is described by:

$$\Delta OD = \frac{1}{2}\left(1 + \mathrm{erf}\left(\frac{\sqrt{4\ln 2}\cdot(t-t_0)}{\tau_p}\right)\right)\sum_{i=1}^{N}\left(A_i \exp\left(-\frac{t-t_0}{\tau_i}\right)\right). \tag{2.12}$$

The amplitude A_i denotes the relative weight of i-th process associated with the decay time τ_i. With the error function one can estimate the time resolution (τ_p) of the measurements.

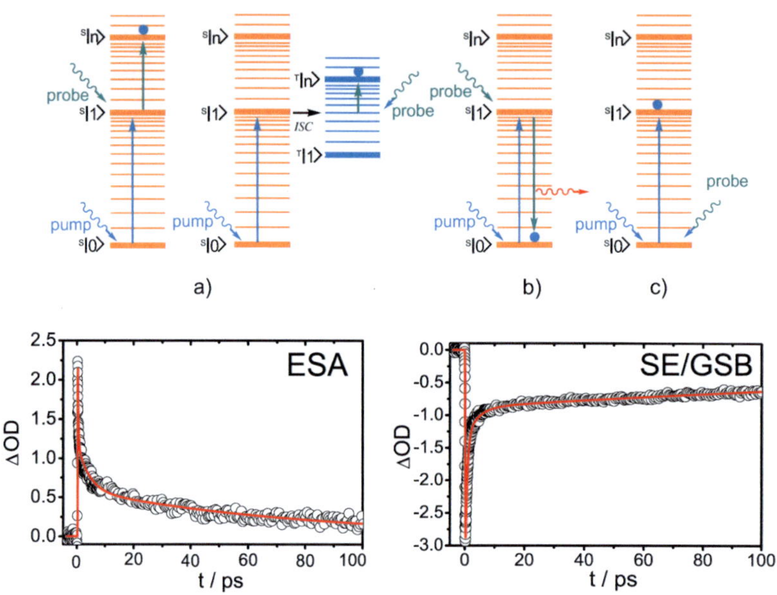

Figure 2.8: Experimental explanations of pump-probe spectroscopy in ultrafast relaxation processes: a) excited state absorption (ESA), b) stimulated emission (SE) and c) ground state bleaching (GSB).

Chapter 3: Ultrafast Dynamics and Transient Anisotropy of Metalloporphyrins

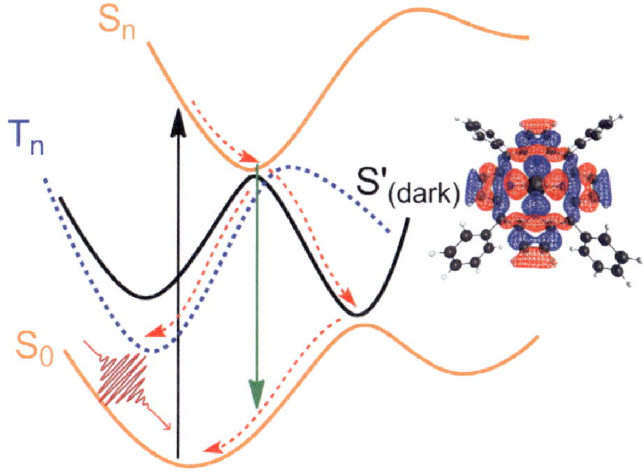

3.1 Introduction

Cyclic tetrapyrroles, and their derivatives, are a class of naturally occurring macrocycles that are ubiquitous natural pigments involved in a wide variety of important biological processes in the world. For example, hemoglobin in heme proteins is responsible for the oxygen transport in the blood[62] while chlorophyll in plant cells is part of the light-harvesting system regulating photosynthesis. Therefore, these molecules remain of fundamental interest to chemists and biochemists. Indeed, they continue to be intensely investigated by many researchers. The basic structure of the porphine is composed of four pyrrole units which are connected to a macrocycle over four methine bridges. Porphine has an aromatic system containing 22 π-electrons, 18 of them are delocalized on the basic structure according to Hückel's rule for aromaticity (see Figure 3.1, left). Its functionality can be tuned by several factors such as the C-N skeleton which can be saturated to a certain degree and bears certain substitutes, and the surrounding, e.g., a protein environment. As a ligand, porphyrin has the ability to form strong metal complexes with a range of cations in the periodic table having incompletely filled d-orbitals (Cu, Zn, Fe, Co, etc.). The interactions take place between the π- and d-manifolds of the porphyrin and the central metal atom, respectively. In recent years, porphyrin-derivatives were investigated for different purposes like light-harvesting complexes in solar cells[63-65], photodynamic therapy in medicine[66-69], and electronic devices[70-72]. Consequently, there exists much interest in the optical properties and the photophysical behavior of porphyrins, especially metalloporphyrins.[73-76] To better understand these properties, model systems such as metal *meso*-tetrakis-phenyl porphyrins (MTPP, M = Mg, Cd) were chosen for investigation.

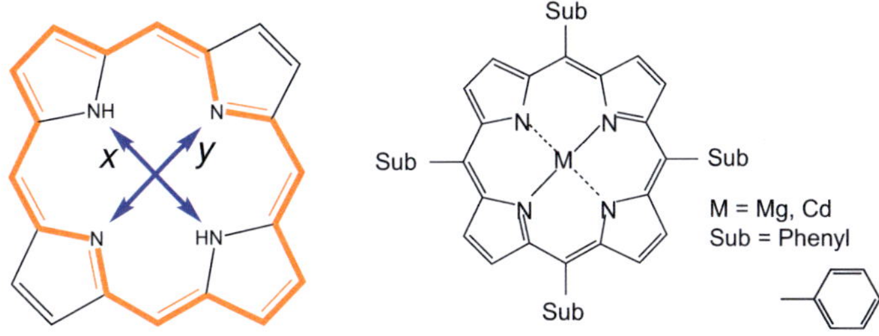

Figure 3.1: Left: the aromatic structure of porphine with an 18 π-electron system. Arrows x and y show the ππ* transition dipole moments. Right: metalloporphyrins containing substitutes and central metal atom.

A simplified molecular structure of metalloporphyrin is presented in Figure 3.1 (right). Typically, its symmetry is supposed to be close to D_{4h} and in the following the D_{4h}-symmetry labels will be used. However, in the minimum energy geometry the four peripheral phenyl rings are known to be twisted about 65-70°[77-79] with respect to the porphyrin skeleton, lowering the symmetry to C_{2v}. In coordinating solvents, Mg atoms in chlorophylls or porphyrins become five- or even six-fold coordinated reducing the symmetry further.[80]

Photophysical properties of many diamagnetic metalloporphyrin derivatives such as ZnTPP, MgTPP and CdTPP have been studied after excitation in the Soret-band. Fluorescence up-conversion experiments by Zewail and co-workers showed that for ZnTPP in benzene, the fluorescence lifetime of the Soret-band was measurably longer than the rise time of the Q-band fluorescence which was interpreted by the presence of an additional state close to the Soret-band.[81] These findings could not be confirmed by other groups,[82-85] however, a small fraction of excited ZnTPP molecules was found to bypass the Q-state.[73,86] In MgTPP it could be shown, that the efficiency for the S_2-S_1 relaxation is close to unity, while it is significantly decreased in CdTPP.[83] Proposed explanations for this behavior are reaction paths via dark *gerade*-states as well as via the triplet-manifold.[82-88] Evidence for the existence of dark states has been given by TDDFT-calculations[89-92] and by direct access upon Q-band excitation and NIR-probe in tetratolyl-derivatives (*meso*-tetratolyl-21H,23H-porphyrin [TTP-H$_2$] and ZnTTP).[93] The role of these states might play an important role in the relaxation pathway of (metallo-)porphyrins. Hence, MgTPP and CdTPP dissolved in tetrahydrofuran (THF) will be studied upon excitation in the Q-band and probing in the visible (VIS) and near infrared (NIR) regions. These experiments allow for the location of dark states up to the N-band. Additionally, TDDFT-calculations were performed on several TPP-derivatives. Based upon the experimental findings and calculations, a model for the relaxation pathway of these molecules upon Soret-excitation is presented, which sheds light on apparent inconsistencies presented so far in the literature.[94]

3.2 Experimental Setup and Conditions

MgTPP and CdTPP were synthesized by the T. S. Balaban group from the University of Marseille (see Ref. [94]). Steady-state absorption spectra of MgTPP and CdTPP in THF were recorded by a Varian Cary5E spectrophotometer and fluorescence spectra by a Cary Eclipse fluorescence spectrophotometer. For time-resolved single color pump-single color probe absorption experiments, a commercial 1 kHz amplified femtosecond Ti:sapphire laser system (CPA 2210, Clark-MXR) with a center wavelength of 775 nm and pulse energies up to 2 *m*J was used. This laser system served as a pump source for two independently tunable NOPA systems (see Appendix A.1). Pump and probe pulses from the NOPA were tunable between 470 and 1700 nm with pulse durations around 25 fs (FWHM) in the visible spectral region using fused silica prisms as an optical compressor. The time-bandwidth product was 0.644 for the pump pulse and 0.531 for the probe pulse at about 620 nm. In this work, the excitation wavelength was set to λ_p = 620 or 571 nm for MgTPP and λ_p = 610 nm for CdTPP with average pulse energies of 0.9 μJ (chopped at 1 kHz). Probe pulses typically had energies of < 0.1 μJ in the visible, and a factor of three less in the NIR region. The intensity for the pump pulse at the location of the sample was 3.6×10^8 W/cm^2. Sample preparation for ultrafast investigations was accomplished by solving MgTPP in THF that was freshly distilled from sodium/benzophenone under vacuum in quartz cells (Hellma) with an optical thickness of 1 mm. It is worth mentioning that some precautions are necessary when performing ultrafast measurements with CdTPP in THF. First of all, photolysis (Xe arc lamp operating at 150 W) and steady-state absorption spectroscopy revealed that CdTPP in THF is unstable upon irradiation (see Appendix B.2). Second, within recording time of a transient at one probe wavelength (ca. 10 min.), the ΔOD-values fluctuated from negative to positive in a standing cuvette (see Appendix B.3). These observations suggest that the molecule undergoes photoreactions upon light exposure. For this reason, measurements were carried out in a flow-cell system. Third, the probe pulses were spectrally confined at 615 nm to roughly 10 nm (FWHM) with pulse duration of about 100 fs in order to record only SE in the absence of ESA in CdTPP after excitation at 610 nm. In order to avoid remarkable aggregation effects, the diluted solution concentration (~5.0×10^{-4} mol/l for MgTPP and ~3.5×10^{-4} mol/l for CdTPP) was adjusted such that the optical density of the sample matched approximately 0.5 at the peak position of the Q(0,0)-band. All experiments were carried out at room temperature.[94]

Measurements of transient anisotropy in CdTPP in THF were carried out with well-determined pump and probe polarization. A Glan-Taylor polarizer (Alphalas) was used to simultaneously monitor the change of the optical density (ΔOD) for the parallel (ΔOD$_\parallel$) and perpendicular (ΔOD$_\perp$) components (see Appendix A.1). Some details for the evaluation of aniso-tropy data are described in Section 3.5.

The white light absorption measurements were performed in the group of Prof. Dr. E. Riedle. In this experimental setup, a commercial 1 kHz Ti:sapphire based laser system (CPA 2001; Clark-MXR) with a center wavelength of 778 nm, 170 fs pulses, and an energy of 1 mJ was used. The pump was generated in a two-stage NOPA system and subsequently compressed with a pair of fused silica prisms down to 20 fs which corresponds to a time-bandwidth product of 0.84.[95,96] The excitation wavelength for both MgTPP and CdTPP was 620 nm. The spectral width was 35 nm. To ensure the same properties for the probe pulse during the entire experiment we delayed the pump pulse with a commercial translation stage and ensured that the spatial overlap and size of the pump stayed constant when moving the translation stage. MgTPP was measured in a 1 mm cuvette from Hellma. The optical density was 0.2, the pump energy 760 nJ, and the pump intensity 60 GW/cm² (6.0 x 10^{10} W/cm²). The measurements of CdTPP were again performed in a flow-through cell. The optical density was 0.3, the pump energy 150 nJ, and the pump intensity 12 GW/cm² (1.2 x 10^{10} W/cm²). All experiments were carried out at room temperature.[94]

3.3 Steady-state Spectroscopic Properties of Porphyrins

In this section, steady-state spectroscopy of porphyrins will be employed to investigate the electronic structure of porphyrins based upon absorption, fluorescence excitation and emission measurements. These photophysical properties of porphyrins play an important role in interpreting the ultrafast dynamics.

3.3.1 UV/Vis Absorption Spectroscopy

The electronic structure of porphyrins in either free base, or metal-coordinated form, is largely characterized by their absorption and fluorescence spectra.[97-101] UV/Vis absorption spectra of MgTPP and CdTPP in THF are shown in Figure 3.2. The intense peak ($\varepsilon = 4.77 \times 10^5$ M^{-1}cm^{-1}) of about 429 nm for MgTPP and 432 nm for CdTPP is called the Soret-band[102] (also B-band or S_2-state) with a weak blue-shifted shoulder assigned to vibrational progression.[103] The peaks of the so-called Q-band (also S_1-state) which has an extinction coefficient of approximately 1.44×10^4 M^{-1}cm^{-1} were observed at 612 (610) nm for the Q(0,0)- and 571 (569) nm for the Q(1,0)-band along with a smaller shoulder at 530 (529) nm for MgTPP (CdTPP). The three Q-band absorption peaks (inset in Figure 3.2) can be understood in terms of vibrational progressions with energy splitting of ~1170 cm^{-1}.[104] The next higher-lying state in MgTPP peaks at 357 nm. The fourth and more prominent absorption band lies around 320 nm and is called the N-band.[105] All these bands have E_u-symmetry and are assigned to $\pi\pi^*$ transitions. These can be qualitatively explained by the well-established Gouterman four-orbital-model[106-108] and by the MO-interpretation[109-111]. Martin Gouterman first proposed his four orbital model in the 1960s to explain the absorption spectra of porphyrins, which arise from π and π^* transition.[112] According to this theory, in a free base porphin system, where two hydrogen atoms are bound to two nitrogens inside the macrocycle, the symmetry is, therefore, reduced to D_{2h}-symmetry in contrast to metallo-porphyrin with D_{4h}-symmetry (see Figure 3.1, right). For free base porphyrin with D_{2h}-symmetry, to simplify, the HOMOs a_u and b_{1u} are relabeled as b_1 and b_2, while the LUMOs b_{3g} and b_{2u} are denoted to c_1 and c_2, respectively (Figure 3.3). The Q-band in free base porphin originates from the transitions $c_1 \leftarrow b_2$ and $c_2 \leftarrow b_2$, while the Soret-band is assigned to the transitions $c_1 \leftarrow b_1$ and $c_2 \leftarrow b_1$ (see Figure 3.3 inset).[97] With this assumption, the intensities of the corresponding B- and Q-band can be represented by

$$B_y = \frac{1}{2}(b_1 c_1 + b_2 c_2) \qquad B_x = \frac{1}{2}(b_1 c_2 + b_2 c_1)$$

$$Q_y = \frac{1}{2}(b_1 c_1 - b_2 c_2) \qquad Q_x = \frac{1}{2}(b_1 c_2 - b_2 c_1). \qquad (3.1)$$

The x and y label represent the orientation of the transition dipole moments showed in Figure 3.1, left.

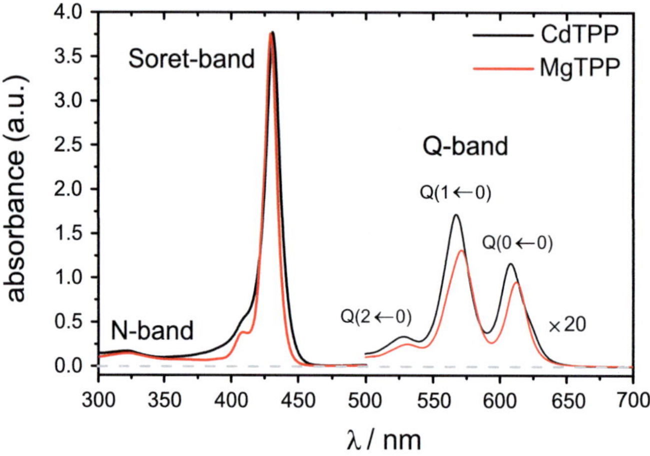

Figure 3.2: UV/Vis absorption spectrum of MgTPP (red) and CdTPP (black) in THF.

In the case of metalloporphyrins with D_{4h}-symmetry, the x and y transitions are equal by symmetry. The Q- and Soret-bands arise from the excitation between the two HOMOs which have a_{2u}- and a_{1u}-symmetry and a pair of e_g-orbitals (LUMOs), that present *gerade* or *ungerade* inversion symmetry. As a consequence, the Q-bands are the result of the transition dipoles canceling each other out and leading therefore to a weak absorbance (1^1E_u-state), while the higher intensive Soret-band (2^1E_u-state) results from a linear combination of the two transitions with higher oscillator strength (Figure 3.4).[97] For metalloporphyrins with different central metal atoms (here Cd and Mg), some small shifts (~ 4 nm, Figure 3.2) on the positions of the bands are due to different interactions of the metal atom with the a_{2u}- and e_g-orbitals.[113] As shown in Figure 3.4, the a_{1u}-orbital has nodes at the pyrrole nitrogen and therefore remains relatively unaffected by the central metal atom.[114] In the MO-interpretation, the electronic structure is explained by the interaction of the four pyrrol rings ($(Py)_4^{2-}$-cage) with CH-methine bridges. For a metalloporphyrin, this leads to several e_g-orbitals and nondegenerate *ungerade*-states. The higher $\pi\pi^*$ transitions in the UV-region are mainly from lower lying *ungerade*-states to the LUMO.

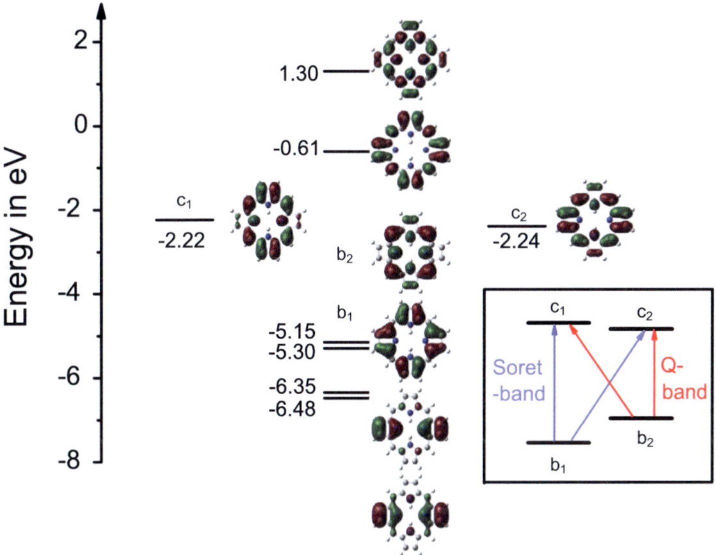

Figure 3.3: The HOMOs b_1, b_2 and the LUMOs c_1, c_2 of free base porphin in the four orbital model (B3LYP/6-31G*).[115,116] Inset shows the transitions between HOMOs (b_1 and b_2) and LUMOs (c_1 and c_2) of free base porphin with D_{2h}-symmetry.

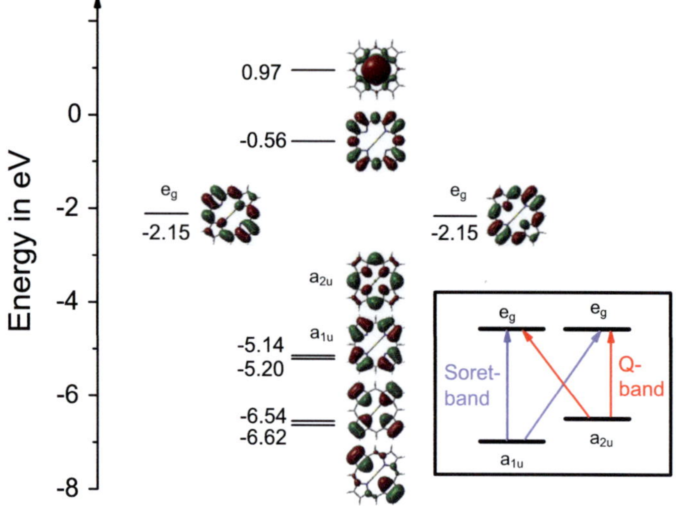

Figure 3.4: The HOMOs with a_{1u}- and a_{2u}-symmetry and the LUMOs in e_g-symmetry of Mg-porphyrin (B3LYP/6-31G*).[117,118] Inset shows the transitions between HOMOs (a_{1u} and a_{2u}) and LUMOs (e_g) of metalloporphyrin with D_{4h}-symmetry.

3.3.2   Solvent Effects

The solvatochromic effect refers to a strong dependence of absorption and emission spectra of a molecule on the solvent polarity.[119] Since polarities of the ground and excited state of a molecule are different, different solvents with varying polarity will lead to a different stabilization of the ground and the excited states of the molecule. Note that the solvent effect on spectra results primarily from the electronic transitions such as $\pi\pi^*$, $n\pi^*$, as well as charge transfer absorptions and depends on the chromophore, namely solvatochromic dyes.[120] Photoexcitation of a molecule leads to electronic density changes in the excited state due to redistribution of charges. This can result in an increase or decrease of the dipole moment of the excited state in comparison with the ground state. The solvatochromic method, which is used to study the electronic and geometrical structure of the molecule, is based on the shifts of absorption and emission maxima in different solvents of varying polarity. With increased solvent polarity, the molecule in the ground state is better stabilized by solvation than the molecule in the excited state, which leads to a spectral blue shift, namely negative solvatochromism; vice versa, better stabilization of the molecule in the first excited state relative to the ground state with increased solvent polarity, will lead to a red shift of the spectra (positive solvatochromism). In order to check the influence of different solvents on the first excited state (Q-band), toluene was chosen for comparison. The UV/Vis absorption spectra of CdTPP in anhydrous toluene and THF are shown in Figure 3.5. It is well known that the relative polarity of THF (0.207) is higher than that of toluene (0.099).[120] With increased solvent polarity, the CdTPP absorption spectra in toluene and THF show a very small hypsochromic (blue) shift of the two Q-bands from 573 to 568 nm and from 615 to 609 nm, respectively (Figure 3.5, right). Since the peaks of the first excited state in THF are blue shifted with respect to that in toluene, this can be assigned to a negative solvatochromism and may reflect a better stability of CdTPP in the ground state.

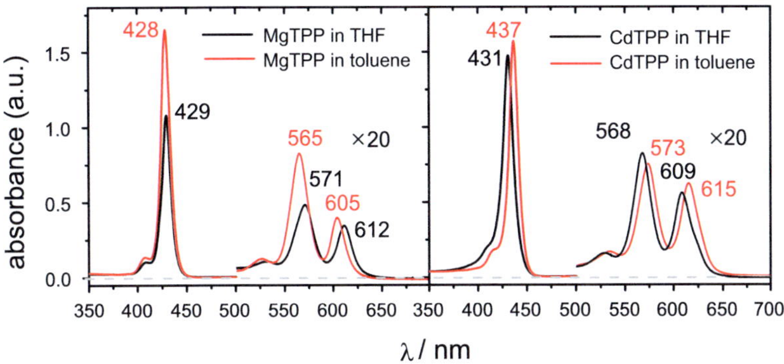

Figure 3.5: UV/Vis absorption spectra of MgTPP and CdTPP in THF (black) and toluene (red), respectively.

In the case of MgTPP, on the contrary, with increasing solvent polarity, the absorption peaks of the first excited state undergo a bathochromic shift (Figure 3.5, left). These observations suggest that THF molecules can

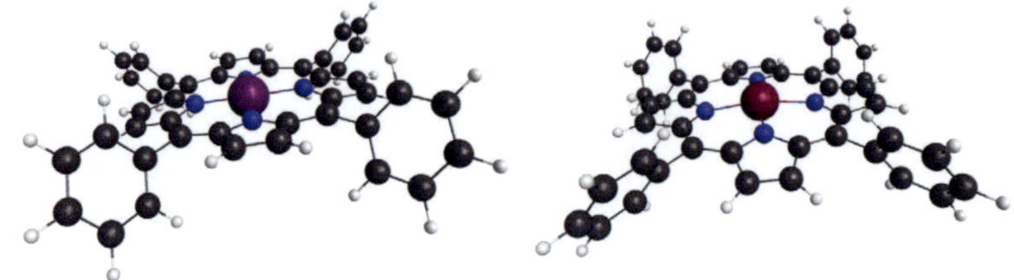

Figure 3.6: Structure of MgTPP (left) and CdTPP (right) using B3LYP/3-21G*.[121]

better stabilize the first excited state of MgTPP than the ground state. TDDFT-calculations of MgTPP and CdTPP at the Franck-Condon geometry show that in contrast to MgTPP, the Cd atom is pushed outside the porphyrin ring and the whole molecule is bent (Figure 3.6).[121] This square-based pyramid structure has been caused by the large size of Cd^{2+} (1.02 Å) and may play a different role of the photophysical properties in porphyrins.[77]

Figure 3.7: MgTPP with one (left) and two (right) THF solvent molecules (B3LYP/3-21G*).[121]

It well known that, in nature, the Mg atom in chlorophylls exhibits almost exclusively five-fold coordination[122-124]. For example, a water molecule can provide an important stabilization in the special chlorophyll pairs P700 and P680.[125] For MgTPP, especially in THF, fivefold coordinated species should be prevalent, beside the six coordinated one[126] (see Figure 3.7). In order to better understand the five- or six-coordinated structure of porphyrins, TDDFT theoretical calculations were performed using B3LYP/3-21G*.[121] From the results of theoretical calculations for MgTPP, coordinated to one or two THF solvent molecules, the position of the phenyl rings, i.e. the angle between the plane of the porphyrin skeleton as well as solvent molecules, play an important role in the spectra of porphyrins. Due to the steric demand of the solvent molecules, the phenyl-rings are supposed to twist considerably which affects the orbital overlap between the skeleton and ring, which exerts influence on the methin-bridges in the MO-model. In the hypothetical case of the phenyl rings lying in the plane of the porphyrin skeleton, the electronic states are lower by 0.4-0.5 eV for the *ungerade*-states on a B3LYP/6-31G*-level. Coordination with one THF molecule leads to an energy decrease of the electronic states and consequently the red shift in the first absorption band (Figure 3.5). Further theoretical studies about the properties of the bond between porphyrins as well as phthalocyanines and central metals without (Mg) or with a closed *d*-shell (Zn and Cd) to ligands binding via various row II and III atoms (e.g. nitrogen, oxygen, phosphorus and sulphur) were performed using energy decomposition analysis (EDA) with various methods and basis sets.[121] The calculation results showed that the ligands binding in metalloporphyrins via a nitrogen atom were preferred to those binding via oxygen. In the case of six-coordinated porphyrin, the phenyl groups are twisted by ~27.7° from perpendicularity and therefore exhibit a different geometry instead of a D_{4h}-symmetry as suggested. Moreover, EDA analysis suggests that six-fold coordination was readily available for magnesium as central metal atom as opposed to zinc and cadmium porphyrins which also showed a weaker bond strength.[121] However, upon the vertical excitation over a rather large laser pulse width, the excited species will be still assumed to be of D_{4h}-symmetry. Further experimental and theoretical work will address this issue. Moreover, four-fold coordinated Mg species within porphyrins and (bacterio)-chlorophylls have never been encountered experimentally in crystal structures which are unstable and coordinately unsaturated. Either solvent or surrounding water molecules are spontaneously tightly bound while in a protein environment, various polar amino acid residues function as apical ligands to the Mg atom.[122] For Cd porphyrins, the X-ray structures of pyridine and toluene adducts of cadmium *meso*-tetra-(p-chlorophenyl)porphyrins have been determined by Wun *et al.*[127] and were proposed as a square-based pyramid.

3.3.3 Fluorescence Emission Spectroscopy

The fluorescence emission spectra of MgTPP and CdTPP have been measured and compared with absorption spectra in THF and toluene, which are displayed in Figure 3.8 and Figure 3.9, respectively. The emission spectra show a mirror-symmetry relationship to the ground state absorption band in the Q-region and exhibit a small Stokes shift which suggests that it undergoes minimal nuclear geometry change after photoexcitation.[128] Besides emission from the Q-band, one also observes fluorescence from the Soret-band violating Kasha's rule (fluorescence spectrum of the Soret-band is not shown here).[129,130] The fluorescence band originates from the first excited state (Q(0,0)) peaks at 617 nm for MgTPP and 626 nm for CdTPP in THF, which is in agreement with the literature[131,132]. The second weak fluorescence band corresponding to the absorption band Q(1,0) was observed at 671 nm for MgTPP (665 nm for CdTPP) on the red side of the first maximum. The $S_1 \rightarrow S_0$ fluorescence spectrum shows a small Stokes shift of roughly 120 cm^{-1} for MgTPP and 434 cm^{-1} for CdTPP. For porphyrins in toluene, the fluorescence spectra have similar behavior as THF. Comparison of the emission spectra between CdTPP and MgTPP show that the emi-

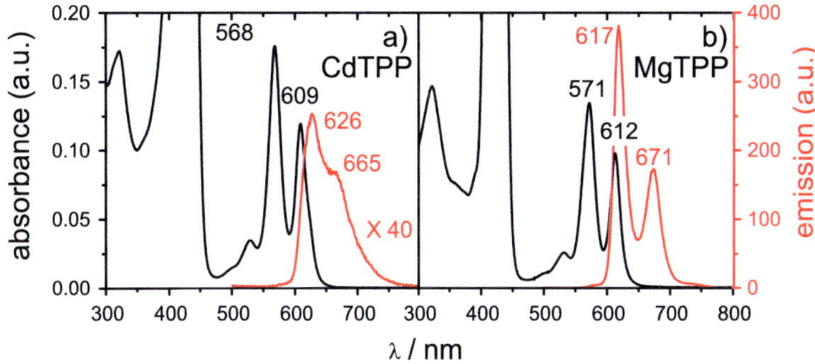

Figure 3.8: Steady-state absorption (black) and emission (red) spectrum of a) CdTPP and b) MgTPP in THF measured at room temperature; emission spectra excited at 431 nm for CdTPP and 430 nm for MgTPP.

ssion spectrum of MgTPP is more structured and the bands are narrower. Moreover, a significant difference was observed for the emission intensities. The fluorescence quantum yield for CdTPP was estimated to be 4 × $10^{-4[132]}$ which is much lower than that for MgTPP (0.15 in Ref. [132] and 0.13 in Ref. [133]). Hence, the fluorescence intensity of CdTPP is much lower than that of MgTPP. This result indicates that the fluorescence properties of both porphyrins are significantly altered with different central metal atoms (here Mg and Cd).

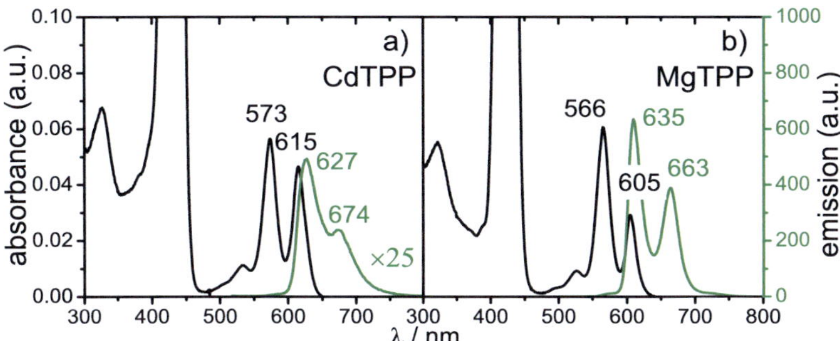

Figure 3.9: Steady-state absorption (black) and emission (green) spectrum of a) CdTPP and b) MgTPP in toluene measured at room temperature; emission spectra excited at 431 nm for CdTPP and 428 nm for MgTPP.

Additionally, concentration-dependent fluorescence measurements were recorded with varying concentrations of CdTPP and MgTPP in THF. From the results as shown in Figure 3.10, it's found that the emission intensity of porphyrins increases with decreasing concentration of porphyrins. In general, this behavior can be interpreted primarily via the formation of aggregates such as dimers and trimers. In an earlier report, decreased emission intensity of the Soret-band was also observed, which suggested that the lower fluorescence quantum yield has been caused by high aggregates of porphyrin in comparison with their monomers.[134] In this case, a Förster-type (fluorescence resonance) energy transfer between molecules leads to reabsorption of an emitted fluorescence photon by neighboring molecules and therefore to reduction in radiative efficiency (low fluorescence intensity). This type of energy transfer requires a similar resonance frequency between energy donor and acceptor. Thus, a small Stokes shift between absorption and emission spectra in porphyrins could give rise to resonance energy transfer. Consequently, this self-assembly process at high concentrations leads to fluorescence self-quenching in porphyrins.

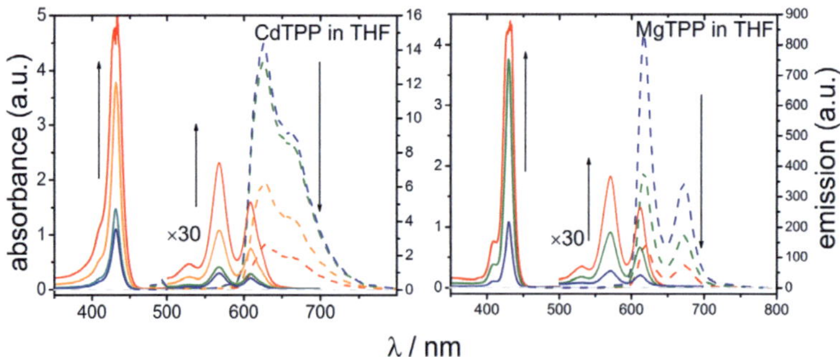

Figure 3.10: Concentration dependent absorption (solid) and fluorescence emission (dash) spectra of CdTPP (left) and MgTPP (right) in THF at varied concentration; red: 1.8×10^{-5}mol/l, orange: 8.3×10^{-6}mol/l, green: 3.1×10^{-6}mol/l, blue: 2.2×10^{-6}mol/l for CdTPP; red: 1.3×10^{-5}mol/l, green: 6.5×10^{-6}mol/l, blue: 2.0×10^{-6}mol/l for MgTPP.

3.3.4 Fluorescence Excitation Spectroscopy

Fluorescence excitation spectroscopy is a powerful technique which enhances the capabilities of fluorescence emission spectroscopy. Unlike fluorescence emission spectroscopy, the fluorescence excitation spectrum is obtained by fixing the emission wavelength and varying the excitation wavelength across a region needed. Principally, equivalent to an absorption spectrum, the fluorescence excitation spectrum characterizes the electronic fine structure of the molecule in the excited states. As shown in Figure 3.11 for CdTPP and 3.12 for MgTPP in THF, the excitation spectra are very similar to the absorption spectra at low concentrations (e.g., 2.0×10^{-6} mol/l for MgTPP) without additional bands (Figure 3.12, blue curve). In the case of high concentrations, the excitation spectra in the Q-band region remain similar to the absorption spectra, while the peaks near the Soret-band region are quite distinct. The Soret band in the absorption spectrum has been split up from 431 nm into 415 and 445 nm, for example, in Cd-TPP. Those observations indicate that at high concentrations ($> 3.0 \times 10^{-6}$ mol/l) the strong interactions in the Soret-band lead to the splitting of the Soret-peak. In contrast, these interactions are very weak for the Q-band. It is well known that porphyrins can self-assemble spontaneously into aggregates through noncovalent interactions depending on the concentration and structural properties.[135-137] This self-assembly leads to fluorescence self-quenching which has been shown in Section 3.3.3. The formation of aggregates can also be envisaged from the fluorescence excitation spectrum. Before interpreting the fluorescence excitation spectrum, a brief overview about the aggregation of fluorescent molecules seems to be necessary. Two classes of aggregates are H-aggregates and J-aggregates. The energy level diagram for a monomer, a H-dimer and a J-dimer has been suggested by Herz[138] and is shown in Figure 3.13. H-aggregates form when the transition dipole moments of two molecules are in a face-to-face configuration (*parallel transitions* dipoles in Ref. [134]), whereas J-aggre-

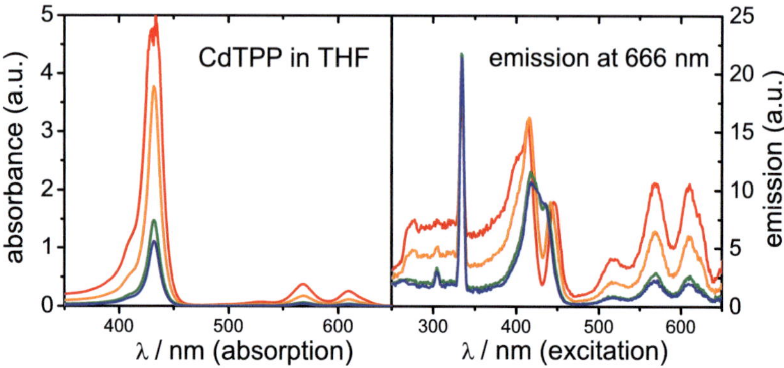

Figure 3.11: Steady-state absorption (left) and fluorescence excitation (right) spectra of CdTPP in THF with varying concentration (1.8×10^{-5} mol/l (black), 8.3 $\times 10^{-6}$ mol/l (red), 3.1×10^{-6} mol/l (blue), 2.2×10^{-6} mol/l (purple)). Excitation wavelengths vary between 250 and 650 nm and the fluorescence is probed at 666 nm.

gates arise when the two transition dipole moments are positioned edge-to-edge (*in-line transition* dipoles in Ref. [134]). In the case of H-aggregates, the out-of-phase dipole interaction leads to a lowering of energy (E'-state), while the in-phase dipole interaction gives rise to repulsion and increases the energy level (E''-state). Since the transition dipole moment is given by the sum of the individual dipole moment of molecules, the transition from the ground state to the E'-state is forbidden, but is allowed to the E''-state. For J-aggregates, again, the in-phase dipole interaction along the polarization axis leads to electrostatic attraction and reduces the excited state energy (E'-state). However, the out-of-phase interaction causes repulsion and produces the higher energetic E''-state. In this case, the transitions to the E'-state are allowed whereas the transitions to the E''-state are forbidden in contrast to H-aggregates (Figure 3.13).

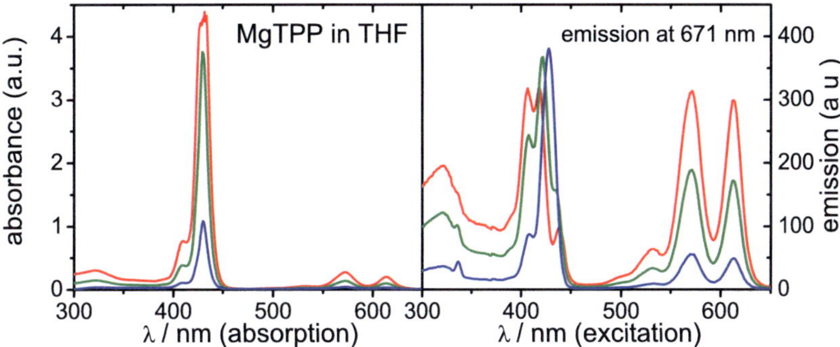

Figure 3.12: Steady-state absorption (left) and fluorescence excitation (right) spectra of MgTPP in THF with varying concentration (1.3×10^{-5} mol/l (black), 6.5×10^{-6} mol/l (red), 2.0×10^{-6} mol/l (blue)). Excitation wavelengths vary between 250 and 650 nm and the fluorescence is probed at 671 nm.

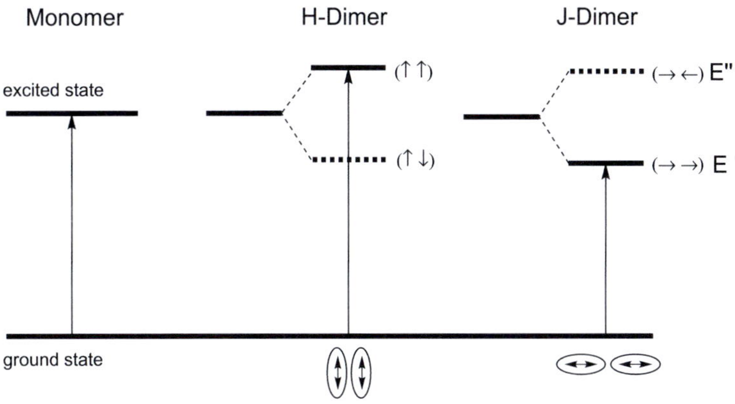

Figure 3.13: Simplified scheme of the absorption from the ground state to the first excited state for a fluorescent molecule for monomer, H-dimer and J-dimer.[134,138].

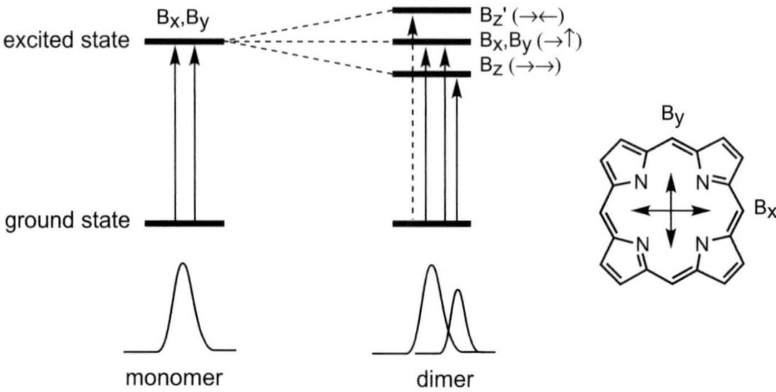

Figure 3.14: The excitonic coupling model in direct linear combination of *meso-meso* linked diporphyrin (illustration adapted from ref. [143]).

As a consequence, in-line transition dipole (J-aggregates) leads to a spectral red shift in comparison with the spectrum of monomer. H-aggregates absorb, however, at shorter wavelength relative to the monomer. In general, J-aggregates are typical for fluorescent molecules, while H-aggregates lead to weak or non-fluorescence.[138] Porphyrins are well suited building blocks to investigate self-assembled processes, since they can spontaneously self-assemble into dimers or higher aggregates through co-valent interactions.[139] From our experimental results of fluorescence exci-tation spectra, the splitting of the Soret-peak from 431 nm to 415 nm and 445 nm may result from the aggregate formation as the concentration of porphyrin increases. According to the excitonic splitting theory[134,140], such a splitting is *in-line transition* (red shift) according to the data from excitation spectra. This simple point-dipole exciton coupling theory is useful to interpret the spectral features caused by the dimeric intera-ctions.[134] At the same time, the peak at 415 nm remains to be assigned to the Soret-transition. This can be interpreted using the excitonic coupling model suggested by Kim and co-workers.[141,142] Figure 3.14 illustrates how the electronic states of the monomer split into several associated electronic states due to excitonic interactions in a *meso-meso* linked diporphyrin.[143] As shown in Figure 3.14, the transition dipole moment of the Soret-band (B_x and B_y, see Eq. 3.1), which is degenerate in a monomeric porphyrin, independently interacts with neighboring porphyrins. The associated electronic states couple to each other and form two excitonic states (B_z' and B_z). B_z'-state is an energetically higher-lying state and dipole-forbidden, whereas the low-lying state (B_z) is allowed. As a results of the splitting, the Soret-band exhibits a red-shifted band and an unperturbed band.[143] In contrast to CdTPP, the splitting of the Soret-band in MgTPP is not very clear (Figure 3.12). At the same time, the Q-bands remain nearly at the same position due to weak oscillator strength and therefore exhibit monomeric features.

3.4 Ultrafast Relaxation Dynamics of the First Excited State

During the past years, ultrafast time-resolved spectroscopy studies have been performed on a number of transition metal porphyrins in order to clarify the dynamics and structures of their excited states.[144-146] Due to the interactions between central atom *d*-orbitals and the nitrogen orbitals of the tetrapyrrole macrocycle, the excited state dynamics depend on the nature of the central metal atom and its significant effects on the excited state lifetimes have been observed and also intensively studied.[81-84,93,132] Thus, excited state characterization is important in determining the possible applications of those compounds. Self-assembly effects at high concentration ($\sim 10^{-4}$ mol/l) which have been mentioned above may not affect the dynamics of the first excited state according to the fluorescence excitation and emission spectroscopy. Thus, in this section, the relaxation dynamics of the first excited state of metalloporphyrins with different central metal atoms (M = Mg, Cd) are studied by means of pump-probe spectroscopy to gain some knowledge about the excited state dynamics and the influence of dark states on the dynamics.

3.4.1   Relaxation Dynamics of MgTPP

In the degenerate experiment with single color pump and single color probe at a wavelength of 620 nm and a spectral pulse with (FWHM) of 25 nm, excitation of MgTPP in THF exclusively showed a positive ΔOD-value, which represents excited-state absorption (ESA), instead of ground state bleaching (GSB) and/or stimulated emission (SE). However, upon reduction of the spectral width to about 7 nm, obtained by two variable thin metal plates mounted inside the optical compressor (Appendix B.1), negative response could be obtained. These spectrally narrow pulses with a pulse duration of about 75 fs (FWHM) were used to probe the ultrafast response in the vicinity of the Q(0,0)-transition between 600 and 630 nm by repositioning the metal plates in the optical compressor while leaving the pump pulse unchanged. The results are shown in Figure 3.15: A negative response is detected at 615 nm and, less pronounced, at 610 nm. In contrast, the pump-induced change of the optical density (ΔOD) becomes positive at 622 nm at delay times exceeding ~4 ps and exhibits no apparent GSB/SE responses at all other probe wavelengths. Fitting with a tri-exponential function delivers three time constants of $\tau_1 = (100 \pm 30)$ fs, $\tau_2 = (3.5 \pm 0.4)$ ps, and τ_3 in the nanosecond timescale (Figure 3.17). However, the amplitude for each temporal profile (the absolute value as well as the sign) strongly depends on the probe wavelength.[94] Note that the depopulation of the excited state is possible through some relaxation channels, for example, internal conversion (IC) to the ground state, fluorescence, or intersystem crossing into one of the triplet excited states (ISC). For MgTPP, the quantum yield was determined by many research groups at about 0.24, 0.29, and 0.47.[147,148] Hence, the remaining two time constants cannot be assigned to electronic relaxation. The time constant on a subpicosecond time scale (τ_1) might arise from polarization effects which are known to be very prominent in MgTPP[149], although measurements were executed under magic angle condition, it is possible that these effects were not completely extinguished. Another, more reasonable, possibility is assumed that the response is caused by fast vibrational relaxation processes in MgTPP in the excited state. The first time constant is only observed in the proximity of the two competing processes, SE and ESA around 620 nm. Here, a geometrical distortion can lead to a dramatic change in the fluorescence intensity at a given probe wavelength (and also, but probably less dominantly, to a change in the ESA properties). The time constant on the picosecond time scale (τ_2) should be also assigned to vibronic relaxation in the Q-state. The dynamics must be within the Q-band because otherwise, no long-time fluorescence would be observed. Moreover, these dynamics cause a long time response which is relatively flat compared to the structured band and should be assigned to the lifetime of the fluorescence. The time constants found in these experiments agree with those found by Galli*et al.*[149], however, they did not assign them to any process.

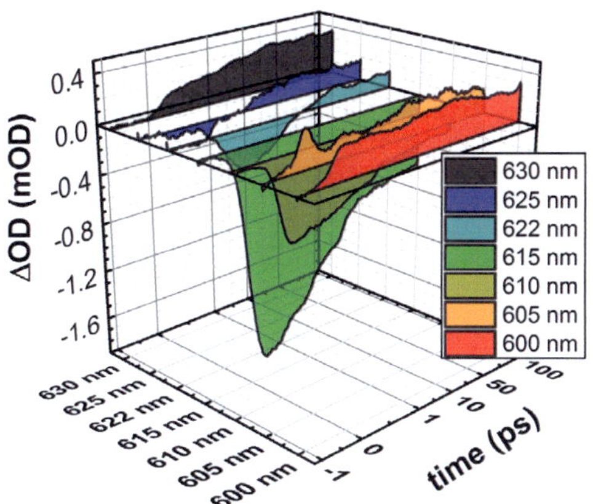

Figure 3.15: Transient responses of MgTPP in THF after excitation at 620 nm and probing between 600 and 630 nm with a spectral width of 7 nm. The time axis is linear up to 1 ps, then logarithmic thereafter.

3.4.2 Relaxation Dynamics of CdTPP

As in MgTPP, there also exists a spectrally confined region of GSB/SE in CdTPP at a probe wavelength between 610 and 615 nm, while a small change in the probe wavelength leads to a dominant ESA contribution (see Figure 3.16). The processes in CdTPP are similar to those in MgTPP. The transient responses of CdTPP can be sufficiently fitted using tri-exponential function and result in three time constants of $\tau_1 = (5.5 \pm 0.6)$ ps, $\tau_2 = (110 \pm 20)$ ps, and τ_3 in nanosecond timescales. The first time constant (τ_1) can be assigned to vibrational relaxation and agree with that in MgTPP. However, the Q-band relaxes on the time scale of 110 ps, indicating the timescale of ISC-process is in agreement with Ref. [132]. The relatively short lived Q-state and the fast ISC-process was also observed and explain-

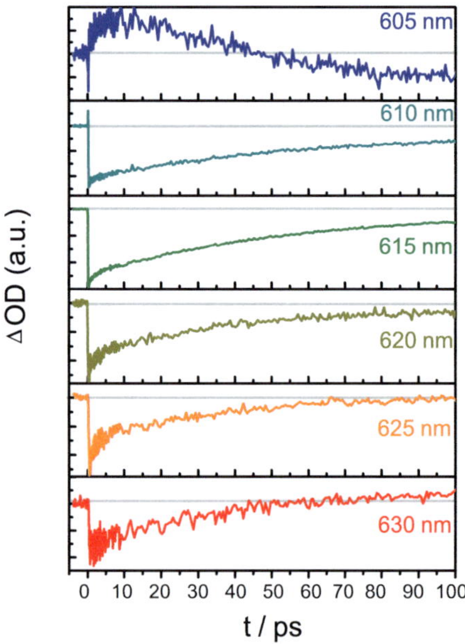

Figure 3.16: Transient responses of CdTPP in THF after excitation at 610 nm and probing between 605 and 630 nm with a spectral width of about 10 nm.

ed in earlier studies.[83] The most significant change is the absence of the fast relaxation (time constant on subpicosecond in MgTPP) in the region around 620 nm (Figure 3.17). One possible explanation is the more rigid structure of CdTPP (Figure 3.6). It is known by X-ray analysis and theoretical calculation that the Cd-atom is pushed outside the porphyrin ring due to the large radius of Cd-atom and the whole molecule is domed.[94,121,150] On the other hand, this destabilizes the molecule but, on the one hand, it is less sensitive to fluctuations and solvent response. Therefore, no shift in the fluorescence spectrum is expected and no response on a subpicosecond time scale could be observed.

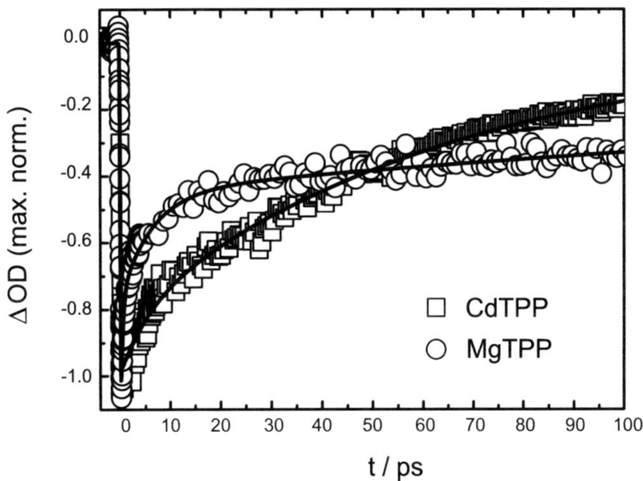

Figure 3.17: Comparison of the negative responses due to GSB/SE of CdTPP (square) and MgTPP (circle) in THF. The pump wavelengths were set at 620 nm for MgTPP and 610 nm for CdTPP, respectively. The probe wavelengths are at 615 nm for both.

3.4.3   The Role of Dark States

The experimental observations (Figure 3.15 for MgTPP and 3.16 for CdTPP) show that vibrational progression can change the sign of the transients from a dominant GSB/SE at short delay times to ESA at longer delays and the GSB/SE response can be obtained only within a narrow spectral window. In other words, spectrally broad pulses lead to an incoherent superposition of both SE and ESA processes, of which the latter process seems to be the dominant part. These observations suggest that the presence of darks states might influence the depopulation of higher excited states. The detection of dark states in the Soret-region is therefore assumed to play a fundamental role in the photodynamics of metalloporphyrins. Although dark states do not contribute to the absorption properties, they might be as crucial for the electronic relaxation as they are in many other systems.[86,151] In a [1+1] excitation process, since two photons are involved, the final states have to be one-photon forbidden *gerade*-states. In subsequent experiments, the probe wavelengths were tuned to the NIR region without confinement of the spectral width of the probe pulse.

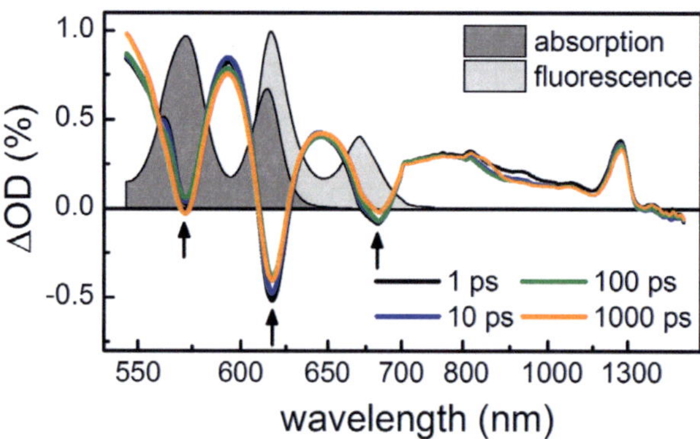

Figure 3.18: Transient absorption spectra of MgTPP in THF after excitation at 620 nm.[94]

46

Figure 3.18 shows the transient spectra of MgTPP in THF performed using a single color pump (FWHM = 40 nm) at 620 nm and a white light probe in the visible region between 450 - 700 nm and NIR reaching up to 1600 nm. Around 600 nm, three bands resulting from GSB/SE can be observed. In the IR, the spectra at 1, 10, 100, and 1000 ps (Figure 3.18) show a broad ESA-band with a maximum at 800 nm (probe wavelength). This band extends even further to 450 nm and clearly dominates the GSB/SE band. Further in the NIR, another noticeable absorption band appears centered at 1240 nm which might be attributed to the weak, quasi-forbidden, Soret-band absorption. All transient profiles showed a mono-exponen-

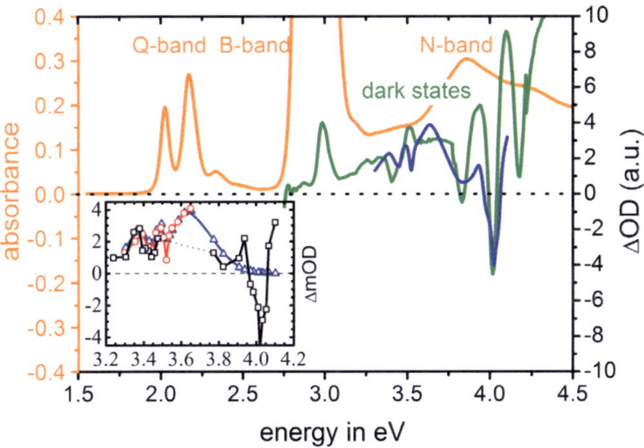

Figure 3.19: Transient spectra in NIR in comparison with the absorption spectrum of MgTPP in THF. Inset shows amplitudes after 620 nm (black) and 571 nm (red) excitation at a delay time of about 1 ps. The spectra were fitted using Gaussian multiple peak function (blue) and resulted in three absorption maxima at 753, 836, and 907 nm corresponding to energy levels of 3.64, 3.48, and 3.36 eV. Dotted line in Inset: masked region for 620 nm excitation due to the limited accessibility of the laser setup. (*x* axis for transient absorption spectra is the combined energy from pump plus probe pulse.)

tial decay with a lifetime in the nanosecond regime. Consider that the broad band around 800 nm might overlap from some different electronic states, additional single 620 nm pump- NIR probe transient absorption measurements were carried out after excitation at 620 and 571 nm corresponding to the Q(0,0)- and the Q(1,0)-band, respectively, and probing in the NIR. The amplitudes at 1 ps of each temporal profile were plotted in Figure 3.19 (inset). Fitted by using Gaussian multiple peak function, the results revealed three maximal absorptions lying between Soret- and N-band at energy levels of 3.64, 3.48, and 3.36 eV which are accessible via [1+1] excitation. The transient absorption spectra and the results of the transient absorptions (NOPA-NOPA experiments) are compared in Figure 3.19 with the steady-state absorption spectrum. The associated data are in good agreement with the results from transient absorption spectra.

The transient spectra of CdTPP in THF probed in the spectral region between 540 and 1700 nm with excitation wavelength at 610 nm is shown in Figure 3.20. In the NIR range, essentially, the spectral features of CdTPP are similar to that of MgTPP: a weak and broad band is going from the visible to 1400 nm with two shallow maxima at 800 and 1000 nm, and an intense band peaks at 1275 nm which is in the same energetic regions as the maximum in MgTPP, and likely has the same origin (Sored-band absorption). However, at least two significant differences can be observed in comparison with the results of MgTPP.[94] First, an additional absorption with a maximum at 850 nm arises beyond 100 ps. According to the proposition by earlier reports[87,88], this absorption band can be assigned to another channel to a triplet excited states via ISC with a decay time constant of 110 ps. Second, the broad ESA-band, which absents in transient absorption spectra beyond 1300 nm in MgTPP (Figure 3.18), could go up to 1600 nm or even longer. Since the amplitudes of the transient absorption spectra are so weak in this range, single color pump-single color probe absorption measurements are required. The experimental results are shown in

48

Figure 3.21. The dynamics of the responses remained almost unchanged at all probe wavelengths. A strong ESA response at time zero was observed around 1000 nm, which corresponds to combined energies of 3.27 eV. In addition, this band seems to extent our accessible spectral window of 1660 nm. The combined energy (pump plus probe pulse) probably falls below 2.78 eV which means that an acceptor state must lie in this energetic region. Further in the visible region the transient absorption spectra of CdTPP differ significantly from those of MgTPP (Figure 3.18). Indeed, two additional ESA bands peak at 550 and 590 nm, respectively, in CdTPP degrading within roughly 100 ps. Afterwards a bleaching is observed that is red-shifted by ~20 nm. Hence, it seems that there is another exit channel in CdTPP on this time scale. Currently, the physical origin of this process is unclear and will be subject to more detailed studies in the future.

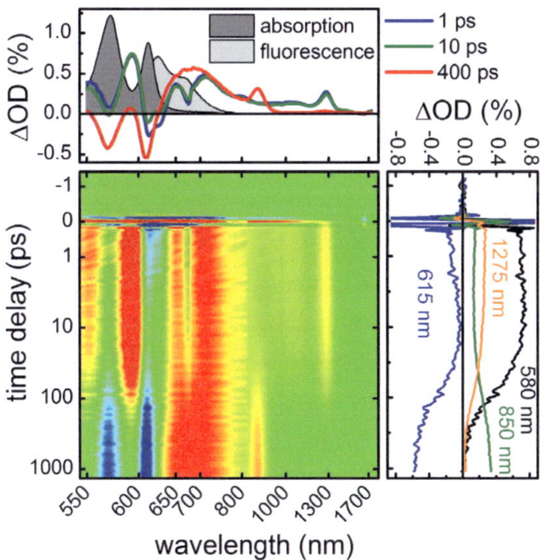

Figure 3.20: Transient absorption spectra of CdTPP in THF after excitation at 610 nm.[94].

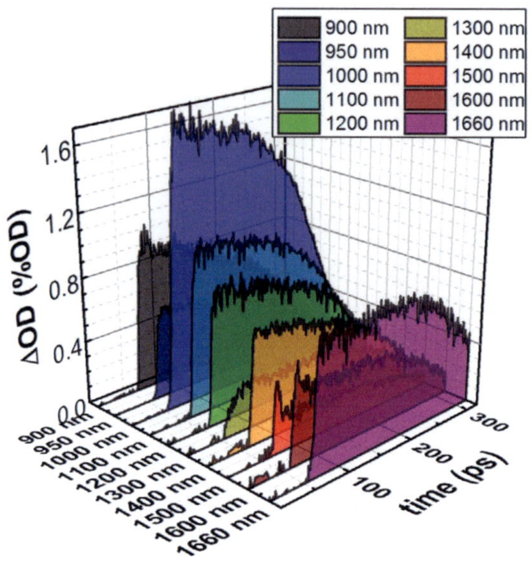

Figure 3.21: Temporal profiles of CdTPP in THF after excitation at 610 nm and probing between 900 and 1660 nm. The time axis is linear up to 1 ps, then logarithmic thereafter.

In order to interpret the experimental findings, TDDFT-calculations are performed by Dr. O. Schalk in the Riedle group on MgTPP and CdTPP using the GAMESS-package and compared with ZnTPP, free base TPP and MgP (magnesium porphyrin without substituent).[94] Ground-state geometries were calculated in B3LYP/6-31G* under the restriction of both D_{4h}- and C_{2v}-symmetry (D_{2d}-symmetry for the case of free base porphyrin). Singlet excitation energies with C_{2v}-symmetry were calculated in B3LYP-/6-31G*. The results are visualized in Figure 3.22. Comparing MgP and MgTPP, one important feature is the energetic difference between equivalent states (about 0.15-0.3 eV). Here the phenyl group in MgTPP plays an important role. These aromatic groups at the methine-bridge allow for a slightly better stabilization than the C-H bond of MgP due to the additional hyper conjugation effect on the porphyrin ring. This effect reduces the interaction between the bridges and the $(Py)_4^{2-}$ cage and gives rise to lower the $5e_g$-state and raise the $4e_g$-state. As a result, the energy gap between $4a_{2u}$-state (HOMO) and $1a_{1u}$-state (HOMO-1) and $5e_g$-state (LUMO) decreases, leading to a red-shift in the absorption spectrum.[97] The differences of the theoretical values for the Q- and Soret-band ($1E_u$ and $2E_u$, see Figure 3.22) between MgP and MgTPP of 0.16 and 0.28 eV, respectively, are in good agreement with the experimental values of 0.11 and 0.23 eV.

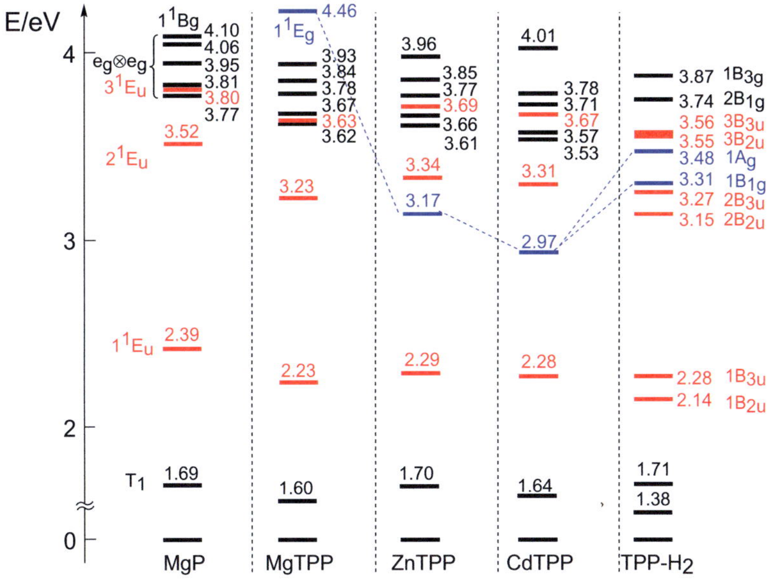

Figure 3.22: Schematic energy diagram for different porphyrin systems calculated on the basis of B3LYP/6-31G* (3-21G* for CdTPP) for bright states (red), dark states (black) and the lowest triplet state (label with T_1). The 1Eg-state is highlighted by blue color and the ground state is in green color.[94]

The excited states of the investigated porphyrins are not very sensitive with respect to the central metal atom. They consist of a set of E_u-states (*ungerade*-states, also bright states) that are one-photon excitation allowed and a bunch of *gerade*-states (also dark states). The E_u-states of the B3LYP are approximately 0.5 eV higher in energy than the experimental values.[94] The *grade*-states consist of 4 states and the $e_g \leftarrow e_g$-excitation between 3.6 and 3.9 eV that cannot be assigned to a certain symmetry, since $e_g \otimes e_g = a_{1g} \otimes a_{2g} \otimes b_{1g} \otimes b_{2g}$ (a_1 and a_2 in C_{2v}-symmetry). The crucial difference among porphyrins with different central metal atoms, however, is an E_g-state resulting from an excitation of a b_{1g}-orbital to the LUMO (the state is highlighted in blue color in Figure 3.22). This orbital is localized at the N-atoms and interacts with an- unoccupied or occupied $d_{x^2-y^2}$-orbital at the central metal (see picture on the cover of this chapter). In the case of the Mg atom, the d-manifolds are empty and this E_g-state can be found around 4.46 eV due to relatively weak interactions. This state is on the blue side of the Soret-band. In ZnTPP, however, the orbital becomes significantly des-tabilized due to interaction of the large size of the $4d_{x^2-y^2}$-orbital as com-pared to the $3d_{x^2-y^2}$-orbital in MgTPP, which causes a poorer overlap with the N-cage. This lowers the gap to the LUMO and reduces the excitation energy to 3.17 eV nearby the Soret-state. The interaction is even more pronounced in CdTPP, where the excitation energy drops to 2.97 eV. In the case of free base porphyrin (TPPH$_2$), the symmetry is reduced from D_{4h} to D_{2h}. This reduction leads to a splitting not only of the Q-band (Q_x and Q_y) and Soret-band (B_x and B_y), but also of the E_g-transition. The latter state turns into a B_{1g}- and A_g-state that are separated by 0.2 eV.[94]

Now one can compare the theoretical results with the experimental data of MgTPP, CdTPP as well as ZnTTP and TTPH$_2$ (porphyrin with tolyl group) from Ref. [93]. The transient spectra of MgTPP, fitted with a Gaussian multiple peak function, reveal three states lying between the optically bright Soret- and N-band at the energy levels of 3.64, 3.48, and

3.36 eV (Figure 3.19). Comparing the energy gaps between these states (~0.15 eV) and those from the vibronic progression of the Q-band, it is difficult to judge whether these states are a vibronic progression within one electronic (dark) state, or if they are correspond to different electronic excited (dark) states. A suitable candidate for those transitions would be states of the $e_g \otimes e_g$-manifold which represent the lowest lying excited *gerade*-states in MgTPP according to these calculations (Figure 3.22 and 3.23).[94]

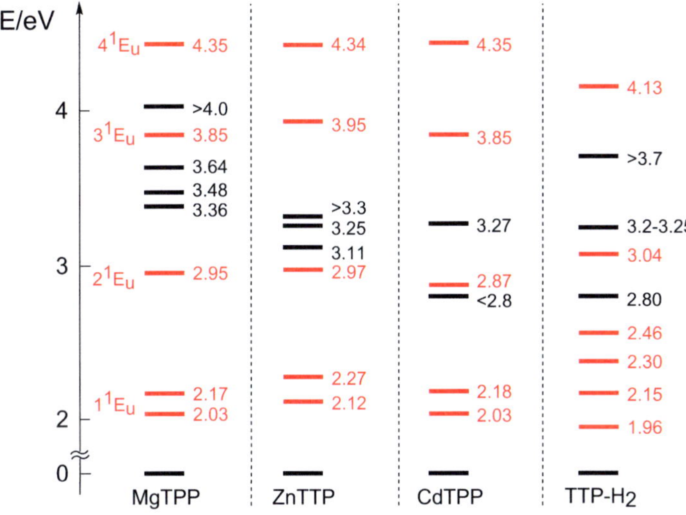

Figure 3.23: Experimental schematic diagram for different porphyrin systems (MgTPP, ZnTTP, CdTPP and TTPH$_2$ in THF). Bright states are in red, dark states in black (data of ZnTTP and TTPH$_2$ are from Ref. [93] and [94]).

In CdTPP, the low lying dark state from calculations at an energy of about 2.97 eV, below the Soret-band, can be supported by the experimental detection at a probe wavelength up to 1660 nm (Figure 3.21). The two weak maxima at probe wavelength around 800 and 1000 nm which correspond to energies at 3.58 and 3.27 eV respectively, arise from the same origin: the $e_g \otimes e_g$-manifold. The position also agrees with the TDDFT-calculations. The transient response at probe wavelength 1660 nm can give the evidence for a further absorption band at a total energy less than 2.78 eV according to the TDDFT-results. The only band that can be responsible for this transient is the 1^1E_g-state which is predicted from DFT calculations.[94]

Previous theoretical and experimental results for the excited states of MgTPP, ZnTTP, CdTPP, and TTPH2 (Figure 3.22 and 3.23) lead to a simple explanation for the relaxation dynamics upon Soret-excitation of these systems.[81-83,88,130] In Mg-porphyrin, the *gerade*-states are all on the blue side of the Soret-band and below the N-band. Those states probably do not influence the relaxation dynamics. Therefore, the molecule exclusively relaxes via the S_2-S_1-channel (from Soret-band to Q-band) which allows the energy gap law to explain the solvent dependence of the lifetime of the Soret-state.[83,152-154] For Zn-porphyrin, the *gerade*-states are nearby the Soret-band and might provide an alternative relaxation channel, possibly bypassing the Q-state. However, the role of this channel is most prominent in Cd-porphyrin, where 30% of the excited molecules are not relaxing via the Q-state[83] and the decay time is significantly shortened. Obviously, the energy decrease of the E_g-state favours its accessibility. In free base porphyrin, the internal conversion is very fast (about 50 fs).[88] These fast dynamics could equally be explained by the presence of an N-localized orbital which was shown to lie below the Soret-band (B_{1g} in D_{2h}-symmetry).[93]

A different way to explain the relaxation dynamics in CdTPP is based on a strong spin-orbital coupling (so-called heavy atom effect[155,156]) which is thought to result in a fast ISC-process to a triplet excited state. This effect was previously used to explain the low fluorescence quantum yield of CdTPP from the Q-state.[132] However, the lifetime of Q-band was determined to be 110 ps in our measurements and the same time constant was previously found upon Soret-band excitation.[132] It is evident that the effect is not sufficient to explain why a large fraction of molecules bypasses the Q-state. Moreover, it does not account for the ultrafast behaviors in TTP-H_2 while the presented model covers both cases. As a conclusion, the experimental observations show that: 1) vibrational progression was able to change the sign of the transient responses from SE at short delay times to ESA at longer delays; 2) SE response can be obtained only within a narrow spectral window for MgTPP and CdTPP. This behavior might arise from the presence of dark states which are one-photon excitation forbidden. The detection of such dark states in the vicinity of the Soret-band is therefore assumed to play an important role in the photodynamics of porphyrin systems. According to the experimental and theoretical results, the presence of dark states below the Soret-band can be used to interpret why the efficiency for the S_2-S_1-relaxation (Soret- to Q-band relaxation) is significantly decreased in CdTPP. The position of the dark states can directly affect the relaxation dynamics upon Soret-excitation since it opens up an additional decay channel.

3.5 Transient Anisotropy

Transient anisotropy[157] is a powerful tool to obtain the knowledge of rotational reorientation[158], intramolecular vibrational redistribution (IVR)[159], and the changes in the electronic transition dipole moment associated to the vibrational relaxation process as well as the rotation diffusion[160]. In addition, it also provides an important contribution to ultrafast molecular dynamics in degenerate systems such as the electronic dephasing in the condensed phase[161]. In a common molecule, excitation occurs along the transition dipole moment which can be regarded as a vector in a three dimensional space. In this case, transient anisotropy has values between 0.4 and -0.2.[162] However, for molecular systems which show two degenerate, or two nearly degenerate excited states, this value can reach as high as 0.7, exceeding the limit of 0.4.[149,163-165] From a theoretical point of view, quantum mechanics[166] and semi-classical models[167] for transient anisotropy in degenerate excited states were developed to describe dephasing and decoherence processes. A direct experimental evidence for such an unusual transient anisotropy values has been obtained by Hochstrasser and co-workers in MgTPP at room temperature[149]. They reported that the transient anisotropy values reached as high as 0.65 at short delay times relaxing to a long time value of approximately 0.1 This behavior was explained by coherent excitation along the two transition dipole moments of the degenerate Q-band[161,168]. In contrast to MgTPP, the transient anisotropy of CdTPP showed an unexpected behaviour where ground state bleaching at very short delay times was followed by excited state absorption for t > 200 fs. The discrepancy between both porphyrins was explained by the presence of "diffuse states" for CdTPP without a more detailed explanation[149]. In order to study the influence of molecular excited states dynamics in more detail, CdTPP in THF was chosen for investigation.[169]

3.5.1 Anisotropy for Nondegenerate and Degenerate States

The anisotropy describes essentially the rotation behaviors of molecules in solution. An incident laser beam with a certain polarization excites only those molecules whose transition dipole moments are parallel to the polarization of the incident light. This is known as photoselection which results in an anisotropic distribution of molecules. Thus, in order to record the change of anisotropy, a second laser beam with polarization either parallel $(I_\parallel)$ or perpendicular $(I_\perp)$ to the excitation beam was employed by means of a tunable $\lambda/2$ plate. A semi-classical approach was developed by Dr. O. Schalk and Dr. A.-N. Unterreiner[167,170] to describe the transient anisotropy in coherent nondegenerate and degenerate system. Here a brief introduction of this model will be presented. In general, the transient anisotropy can be defined as[171]

$$r(t) = \frac{I_\parallel(t) - I_\perp(t)}{I_\parallel(t) + 2I_\perp(t)} \tag{3.2}$$

where

$$I_\parallel = C \langle \cos^{2m} \xi \cos^{2n} \vartheta \rangle_{\varphi,\vartheta} \equiv \frac{C}{4\pi} \int_0^{2\pi} d\varphi \int_0^\pi d\vartheta \cos^{2m} \xi \cos^{2n} \vartheta \sin \vartheta \tag{3.3}$$

$$I_\perp = C \langle \sin^{2m} \xi \sin^{2m} \varphi \cos^{2n} \vartheta \rangle_{\varphi,\vartheta}$$

$$\equiv \frac{C}{4\pi} \int_0^{2\pi} d\varphi \sin^{2m} \varphi \int_0^\pi d\vartheta \sin^{2m} \xi \cos^{2n} \vartheta \sin \vartheta. \tag{3.4}$$

Here m and n are the number of pump and probe photons, respectively. C is constant and can be consider as 1 to simplify the solution. The angles ϑ and φ denote the angles between z- or y-axis and the projection of the transition dipole moment $\vec{\mu}_1$ (here excitation pulses) in spherical coordinates, respectively. ξ is the angle between the second transition dipole moment $\vec{\mu}_2$ (probe pulses) and z-axis. The two dipole moments have an angle of β. Angle γ defines the position of a grand circle that is determined by the two transition dipole moments (see Figure 3.24). After integrating Eq. (3.3) and (3.4) by considering one photoexcitation ($m = 1$) and using Eq. (3.2), the transient anisotropy at time zero results in

$$r(0) = \frac{2n}{2n+3} (3 \cos^2 \beta - 1). \tag{3.5}$$

In the simplest case, the pump and probe photons are absorbed via a one photon process, namely [1+1] process ($n = 1$), along two parallel oriented transition dipole moments $\vec{\mu}_1$ and $\vec{\mu}_2$. In this case, one gets the well-known Perrin equation.[172]

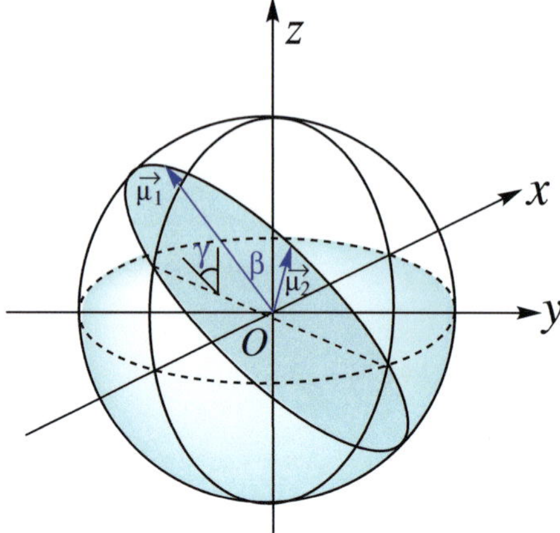

Figure 3.24: Simplified scheme of the transient dipole moments and the relevant angles.[170]

As mentioned above, the angle β describes the polarization direction between the pump induced transition dipole and the probe transition dipole. For the polarization of probe pulses which are parallel ($\beta = 0°$) or perpendicular ($\beta = 90°$) with respect to that of the pump pulse, one obtains zero time anisotropy $r(0) = 0.4$ and $r(0) = -0.2$, respectively. In Eq. 3.5, the magic angle condition ($r(t) = 0$ by $\beta = 54.74°$) is established for common pump-probe experiments in order to avoid the anisotropic effect.

In a system with two-fold degenerate excited states ($|a\rangle$ and $|b\rangle$), both dipole moments are perpendicular to each other. One of them is parallel to the excitation dipole moment. The transition dipole moment $\vec{\mu}_2$ is linear combination of $\vec{\mu}_a$ and $\vec{\mu}_b$

$$\vec{\mu}_2 = a\vec{\mu}_a + b\vec{\mu}_b \quad \text{with} \quad a^2 + b^2 = 1. \tag{3.6}$$

Calculation $I_\parallel$ and $I_\perp$ with excitation along transition dipole moment with maximal projection onto z-axis (in this case $\beta = \beta_0$) results

$$I_{\parallel} = \frac{1}{4\pi} \int_0^{2\pi} d\varphi \int_0^{\pi} d\vartheta \left(\frac{1}{2\pi} \int_0^{2\pi} d\gamma \, (\cos\beta_0 \cos\vartheta - \sin\beta_0 \sin\vartheta \cos\gamma)^2 \right)$$

$$\times \cos^2\vartheta \sin\vartheta = \frac{8}{30} \tag{3.7}$$

$$I_{\perp} = \frac{1}{4\pi} \int_0^{2\pi} d\varphi \cos^2\varphi \int_0^{\pi} d\vartheta \, (\frac{1}{2\pi} \int_0^{2\pi} d\gamma$$

$$(1 - (\cos\beta_0 \cos\vartheta - \sin\beta_0 \sin\vartheta \cos\gamma)^2)) \times \cos^2\vartheta \sin\vartheta = \frac{1}{30} \tag{3.8}$$

$$\text{with } \beta_0 = -\arctan(\cos\gamma \tan\vartheta). \tag{3.9}$$

After using Eq. (3.2) we obtain the zero time anisotropy in $r(0) = 0.7$.[167]

3.5.2 Anisotropy for CdTPP in THF

Figure 3.25a shows the experimental results for three different polarizations (parallel, perpendicular, and magic angle). The experimental anisotropy calculated from the transients of Figure 3.25a using Eq. (3.2) is plotted in Figure 3.25b. An unusual high transient anisotropy up to 0.65 in CdTPP was observed which decays within 100 ps. The dye Nile blue was chosen as reference which showed anisotropy values between 0.35 and 0.4 in the corresponding probe wavelength region. As mentioned in Section 3.4.2, outside the probe wavelength regime of 610 nm $< \lambda_{pr} <$ 620 nm, a small change of the probe wavelength leads to a complex superposition of ESA, ground state bleach (GSB), and SE which is represented by an alternation of negative and positive ΔOD-values. In essence, this superposition is caused by the presence of dark state (*gerade*-states in D_{4h}-symmetry, more details see Section 3.4.3) which are accessible via [1+1] excitation. For most combinations of pump- and probe pulses, the initial anisotropy $r(0)$ is smaller than 0.4. This suggests that the high transient anisotropy value can only be observed when the spectral bandwidth of the probe laser is carefully adapted to the regime of stimulated emission (see Figure 3.16 and 3.27).[169]

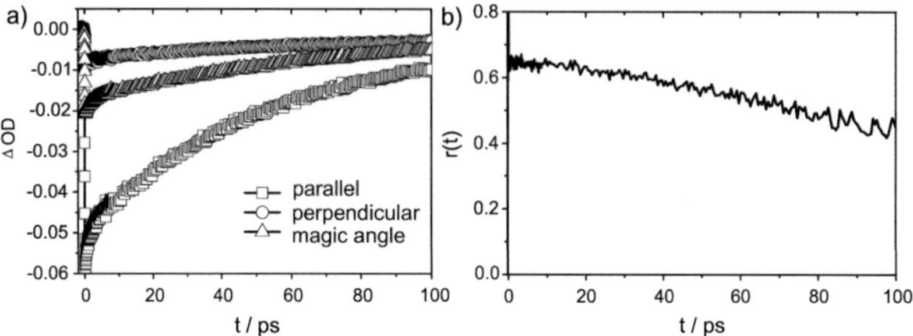

Figure 3.25: a) Transients in parallel (square), magic angle (triangle), and perpendicular (circle) polarizations after excitation at 610 nm and probing at 615 nm. B) Calculated transient anisotropy of CdTPP in THF using Eq. (3.2).[169].

The transient anisotropy of CdTPP in THF in Figure 3.25b decays from an initial value $r(0) \sim 0.65$ in the order of 100 ps.[169] In contrast, the transient anisotropy of MgTPP in THF decays bi-exponentially from about 0.7 within 2 ps to a long time value of approximately 0.1 (see Figure 3.26).[161] Note that for coherent excitations, such as claimed for MgTPP, the anisotropy decay is usually on the scale of a few ps or faster. Since the high transient anisotropy arises from the initial superposition of the two degenerate states (here Q_x and Q_y states), according to theoretical calculations of the molecular structures (see Figure 3.6), the Cd atom is pushed slightly outside the porphyrin ring and the whole molecule is bent due to the relative large radius of Cd atom in contrast to MgTPP. But the whole molecule is still considered as a quasi-degenerate system allowing for a high anisotropy. The restricted flexibility (no fluctuation of the Cd atom in and out of the porphyrin plane) may also be the reason for the much longer anisotropy (dephasing) decay. In other words, the slow decay could be attributed to the specific structure of CdTPP which is more rigid. On the other hand, this structural stiffness may favour instability of the molecule upon irradiation leading to consecutive reactions on a much longer timescale.

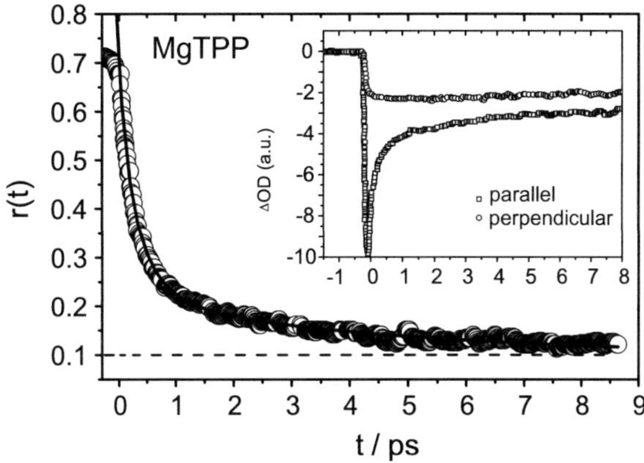

Figure 3.26: Transient anisotropy of MgTPP in THF after degenerate excitation and probing at 620 nm.[161].

3.5.3 Simulation of the Transient Anisotropy

The high initial transient anisotropy, and long decay timescale, could be that they are attributed to the superposition of transient responses. As mentioned above, the negative ΔOD-values might contain an incoherent superposition of GSB/ESA responses. It was demonstrated that these superpositions can influence transient anisotropy such that unusual high values can be observed (see Figure 3.27 and also Ref.[173]). In this case, a simulation of the experimental results is necessary under consideration, in which the observed ΔOD-values are given by the ΔOD-values of ESA components plus ΔOD-values of GSB/SE components

$$\Delta OD_{\parallel/\perp}(Exp.) = \Delta OD_{\parallel/\perp}(ESA) + \Delta OD_{\parallel/\perp}(GSB/SE). \qquad (3.10)$$

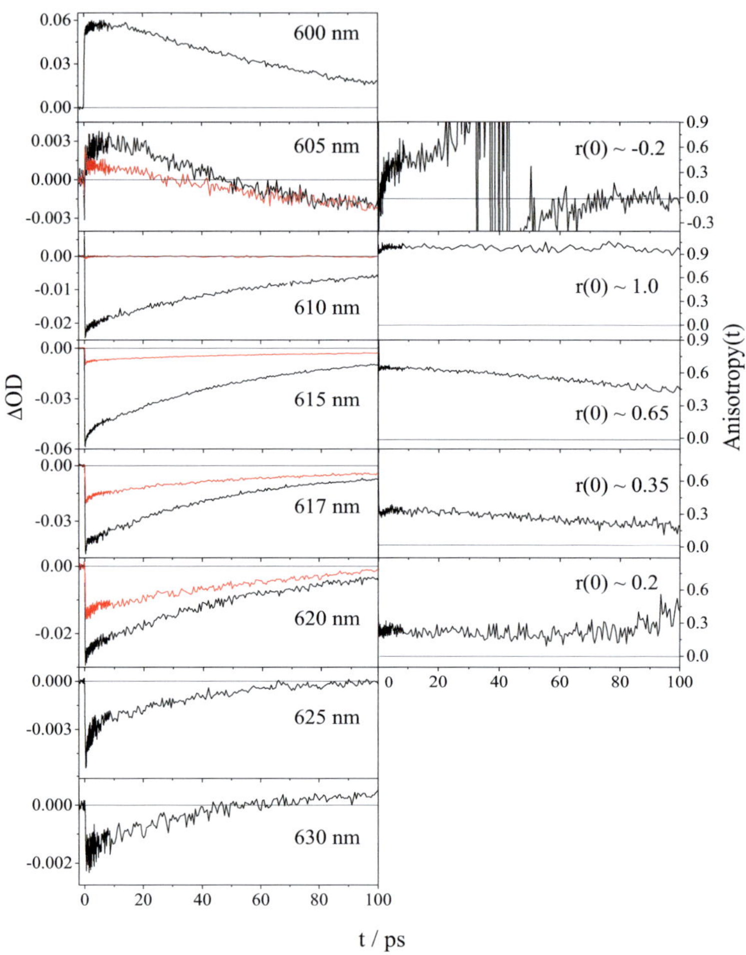

Figure 3.27: Probe wavelength dependence of transients and corresponding transient anisotropy of CdTPP in THF after 610 nm excitation of the Q(0,0)-band. The pump-probe transients with the probe polarization parallel (black) and perpendicular (red) to the pump polarization.[169]

In Eq. (3.10) one assumes that only one dark state was taken into account in the simulation although it is likely that more low lying dark states are available in CdTPP (see Section 3.4.3). The simulation results are shown in Figure 3.28 and suggest that this is possible for those cases where ΔOD caused by dark states (about -0.2) is almost non-polarization dependent (Figure 3.28, left) whereas the bleaching (GSB/SE) contributions contain an "ordinary" anisotropy of roughly 0.4 (Figure 3.28, middle). The two anisotropy components of ESA and GSB/SE lead to the unusually high transient anisotropy which is observed in the experiment. The high symmetry favors these findings through a defined structure of *gerade-* and *ungerade-* states in porphyrins. The time dependence of the anisotropy on the fs-time-scale is governed by rearrangements in the excited state which can have drastic variations given how small geometry changes can interchange between GSB, SE and ESA as observed in the wavelength dependence of the zero time anisotropies (see Figure 3.27). Moreover, the long-time constant about 110 ps shows that about 30% excited molecules can undergo direct intersystem crossing to triplet manifold in CdTPP and suggest an additional relaxation pathway competing with ground state bleaching.[83]

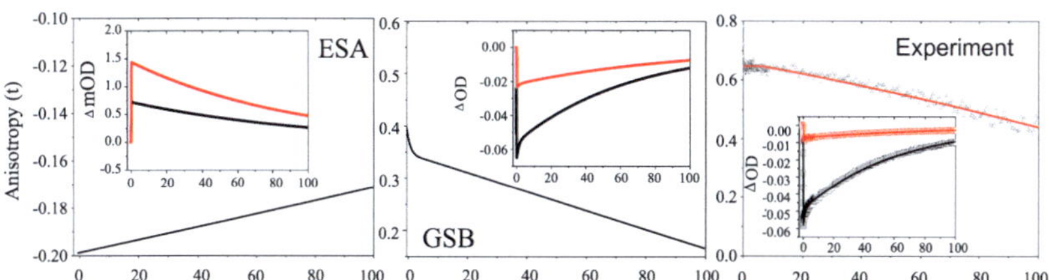

Figure 3.28: Simulated of transient anisotropy after excitation at 610 nm and probing at 615 nm. The inset shows the corresponding transient responses.[169]

3.6 Conclusion and Outlook

The ultrafast relaxation dynamics following the Q-band excitation and transient anisotropy of metalloporphyrins (MgTPP and CdTPP) in THF solution were studied using time-resolved pump-probe spectroscopy. The experimental results and *ab initio* calculation revealed that a relaxation channel via a dark state might account for discrepancies found previously in systems with heavier central metal atoms. This dark state originates from an orbital localized on the central N-atoms and continuously drops in the series of Mg, Zn, Cd. In addition, a fast and efficient intersystem crossing in CdTPP was first direct observed and determined to be 110 ps. Furthermore, an unusual high transient anisotropy up to 0.65 was obtained which decays on a 100 ps timescale. This observation was explained using simulation method by an incoherent mixing of ESA, GSB, as well as SE. Since most porphyrins and corresponding derivatives exhibit five-, or six-fold coordinated complexes in nature, the exploitation of coordination or electrostatic interaction between the axial ligands and the *meso* positions of porphyrinic system seems to be more interesting. Therefore, for a complete understanding of the photophysical and photochemical properties in porphyrinic systems, further study will be focused on the ultrafast dynamics of these five- or six-fold coordinated porphyrins and derivatives.

Chapter 4: DNA-mediated Charge Transfer

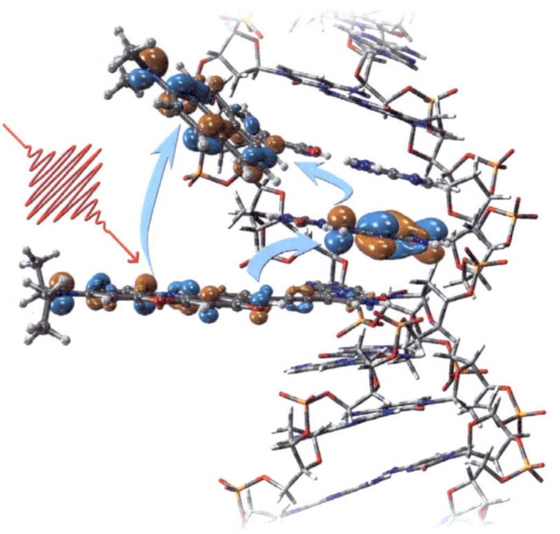

4.1 Introduction

Since the discovery of the first three-dimensional double helical structure of double stranded deoxyribonucleic acid (DNA) by Watson and Crick in 1953 with the aid of crystallographic images[174,175], the structurally well-defined π-stack of DNA has been suggested as a novel medium in which to explore π- stack mediated electron or charge transfer.[176] From a structural point of view, a single DNA strand contains a polyanionic sugar-phosphate backbone connecting four DNA bases, adenine (A), guanine (G), cytosine (C), and thymine (T). The bases cytosine and thymine are derivatives of pyrimidine which contain only a single aromatic ring whereas the bases adenine and guanine are derivatives of purine which has two fused aromatic rings (see Figure 4.1, middle). The double helix DNA is formed when two single strands combine exactly to form extended π- stacked arrays of A:T and C:G base pairs, which are stabilized by very effective hydrogen bonding between these complementary bases of each strand (Figure 4.1, left). The Watson-Crick C:G base pairs possess three hydrogen bonds, whereas the A:T base pairs connect with two hydrogen bonds. Each hydrogen bond contributes approximately 2 kcal/mol energy to the stability of the DNA double helix, meanwhile the distance between two base pairs is about 3.4 Å (namely 1 base pair).[177] Those hydrogen bonds (N—H···O and N—H···N) are the most flexible part in the DNA structure. When a charge, or electron, traps or locates to a nucleoside in the DNA strand, the surrounding hydrogen-bonded protons have to shift in order to "fit" the new charge distribution. The energy barrier that impedes such changing is very low, so that it is hard to accurately determine the location of a trapped charge, in other words, this proton shift is relatively free. Spectroscopic detection of the existence of charge on the bases is difficult since the common bases A, G, C, and T have very weak absorption in the visible region. Due to the shift of a proton, charge migrations through the DNA double helix can take place.[178] Generally, electron/charge transfer processes in DNA are relevant to radiation damage of DNA, since the nucleic acids can be reduced/oxidized by chemical and photochemical methods.[179] From this point of view, π-stack arrays can provide particularly a unique medium for electron or charge migration and even more play important role for the repair of damaged DNA. Consequently, on the one hand, understanding charge migration is thus essential to understand oxidative damage to DNA, which is believed to be partially responsible for aging, apoptosis, and cancer.[180-182] On the other hand, the properties of charge transfer through DNA could be used as a molecular wire for electrochemical biosensor and nanoelectronic devices.[183-185] This idea first came from Eley and Spivey in 1962 based on their conductivity studies in dry samples of DNA and RNA.[186] After that, most studies of DNA-mediated charge transfer have been performed using B-form DNA[187], due to their energetically favorable thermodynamic stability.

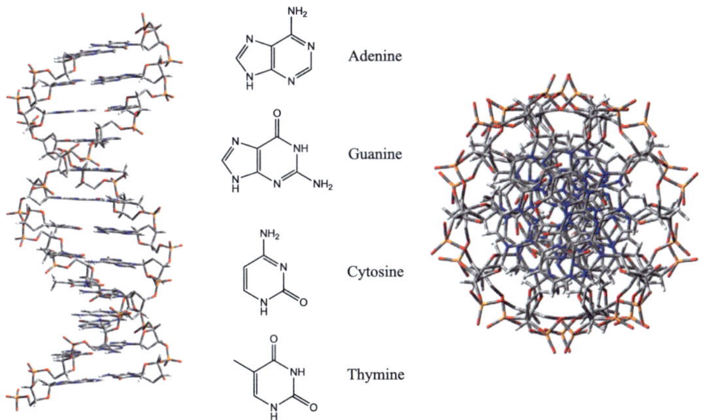

Figure 4.1: Left: structure of a B-from DNA; Middle: the molecular structure of the four naturally occurring nucleotides adenine (A), guanine (G), cytosine (C), and thymine (T); Right: top view of a DNA double helix.

As mentioned above, photoinduced charge transfer in a donor-DNA-acceptor system such as semiconductor[188], organic and biological sensors[189] has been intensely studied in the last decades. Typically, in contrast to electron transfer which can be described as an excess electron propagating through DNA[190], charge transfer through duplex DNA is the migration of charge carried by a radical cation, namely a "hole". Hole migration through DNA has been shown to travel over long distances. This transfer mechanism could only be explained by a charge hopping model with guanine radical cations as intermediate charge carriers. For longer ranges without guanine, hole hopping occurs via adenine radical cations.[191] On the other hand, it is now consensus that short-range hole transfer occurs coherently via a Marcus-type superexchange mechanism that shows an exponential distance dependence of the charge transfer rate. According to chemical[192] spectroscopic experiments[193] as well as theoretical results[194], the turning point between the two mechanistic regimes is established at charge transfer distances of about 3-4 base pairs. However, although these two types of transfer mechanisms have been mostly discussed in detail and widely accepted, the transfer rates are still under debate.[195-198] Moreover, in contrast to the hopping model, the charge transfer mechanism over short-ranges exhibits a more complicated character. Note that the charge transfer rates depend on temperature, DNA sequence, type of charge donor and acceptor, as well as surrounding conditions.[195,199-203] From this perspective, the underlying mechanism for charge transfer over short-ranges in DNA is still unclear. Indeed, from a theoretical point of view, charge hopping could occur potentially also over short distances in DNA.[204-206] Nevertheless, the issue of parallel and possible competing channels for charge transfer between donor and acceptor has drawn renewed attention. In fact, multiple processes involving superexchange and hopping mechanisms on the basis of the spectroscopic and theoretical studies at short-ranges have been suggested.[191,207-209]

Generally, most studies on charge transfer through DNA reported were based upon the observation of fluorescence quenching of charge donors, which were directly attached to the DNA strand. The presence of a charge acceptor usually leads to a fluorescence decrease of the donor and provides a driving force for efficient quenching of the donor singlet excited state. The quenching efficiency was usually assumed to be dependent on the distance between donor and acceptor and, therefore, may be associated with the charge transfer rate over a short-range. Although it is quite clear that the quenching efficiency is related to charge transfer rates, detailed investigations of such correlation has drawn relatively little attention. Furthermore, to understand this correlation may lead to better understanding the superexchange mechanism over a short-range charge transfer. In this work, the attention will be focused on charge transfer mechanisms and rates over short-ranges as well as the correlation between fluorescence quenching efficiency and the timescales using time-resolved pump-probe absorption spectroscopy. [210]

4.2 Charge Transfer Mechanisms and Rates

The mechanisms of charge transfer through DNA were subject of intense study over the last two decades. Despite the large amount of work done in this area, it still attracts the attention of both theoreticians and experimentalists and leads to a heavy debate on the rates and mechanisms of charge transfer.[192,211-217] Generally, for photoinduced charge transfer (CT) in a donor-bridge-acceptor system, the role of the molecular bridge must be taken into consideration. From this point of view, two types of charge migration processes can be identified depending on the relative energies of the donor, bridge, and acceptor levels.[218-220]

4.2.1 Single-step Superexchange

One of the charge transfer mechanisms is coherent tunneling, also called as one single-step superexchange mechanism, in which the charge tunnels from the charge donor to the acceptor without bridge reduction or oxidation. This mechanism therefore requires the donor and acceptor to be energetically well separated from the bridge states, such that the intervening base pairs do not act as charge carriers (see Figure 4.2 a).[221-223] Because the strength of the electronic interactions between donor and acceptor decreases with the increasing number of base pairs separating both redox partners, the rate of charge transfer drops rapidly with long distances. According to the basic Marcus theory, the charge transfer rates of superexchange process are characterized by an exponential donor acceptor distance:

$$k_{CT} \propto exp(-\beta R). \qquad (4.1)$$

where $\beta = 0.6 - 1.4\,\text{Å}^{-1}$ for a DNA bridge, allowing only a short range transfer. In this model, the electronic coupling depends exponentially on the distance between donor and acceptor and, therefore, may dominate the transfer rate. In this case, the distance decay constant β can well describe the correlation between the exponential decrease of the charge transfer rate and the increasing donor-acceptor distance. Jortner *et al.*[218] in their earlier studies suggested that for large β values, the charge transfer can take place as single-step superexchange. To account for small β values, the hopping model has been proposed. More details about β values in the case of charge transfer in DNA were reported in Ref. [218] and [224]. R is the distance between the donor and acceptor.

4.2.2 Multistep Hopping

The other mechanism, called the incoherent hopping mechanism, is charge multistep hopping through the appropriate nucleobases on the bridge. In contrast to superexchange, a charge in the multistep hopping mechanism is first injected into the bridge via oxidizing or reducing the bridge, and generating a new intermediate state (D^*-B^+-A) with low HOMO energy. The charge hops between localized base pairs randomly until it reaches the acceptor. In order to hop from one bridge site to another, the charge requires thermal activation to overcome the energy gap between intermediate and bridge states. This is why the incoherent hopping mechanism is usually called thermally activated hopping. The rate constant of multistep hopping depends weakly on the distance which was observed experimentally by Giese's group[225] and can be described by

$$k_{CT} \propto N^{-\eta} \tag{4.2}$$

where N is the number of hopping steps and η exhibits a value between 1 and 2. Therefore, unlike the coherent single-step superexchange, incoherent hopping in DNA involves several steps, i.e., injection of the charge into the first base of the DNA, the charge transportation along the π-stack due to successive transitions between the nucleobases with the appropriate energetics, and/or charge trapping in the position of the charge acceptor. However, each single step of the hopping model is still considered as an individual superexchange process on the basis of Marcus hypothesis.

In the hopping model, the possible hopping carriers are the four different nucleobases, i.e., adenine, guanine, cytosine, and thymine. These bases have different ionization potentials, which give rise to varied bridge states in different energy levels. The ionization potential of nucleotides increases in the order of G (1.47) < A (1.96) < C~T (2.1) (V versus NHE).[226] Since guanine has the lowest ionization potential, charge transfer takes place predominantly via hopping between guanine sites. As a consequence, oxidation of DNA leads to the formation of a G radical cation ($G^{\cdot+}$) in DNA. As mentioned in the introduction, charge migration through a DNA strand is efficient and almost distance-independent. Recently, it was suggested experimentally that adenine can act as a charge carrier and play an important role as intermediated hole transfer.[191,227] Indeed, recent theoretical calculations suggest that hole transfer through A:T base pairs may be more favorable.[228] In contrast, thymine and cytosine have the highest electron affinity and are most easily reduced. An excess electron injected into DNA and migrated via thymine or cytosine radical anions will therefore trigger excess electron transfer through DNA.[197]

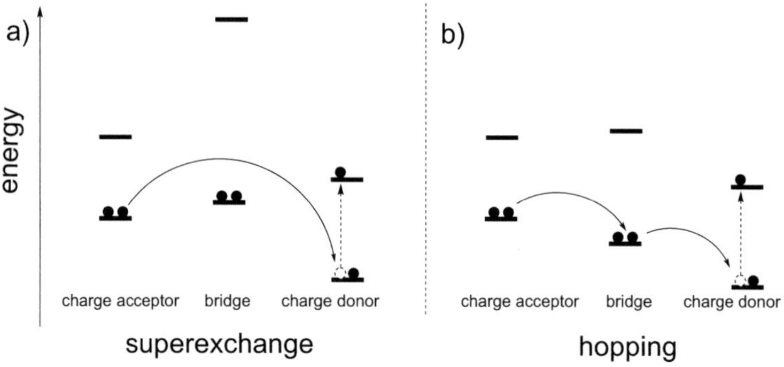

Figure 4.2: Schematic representation of a) superexchange and b) hopping mechanisms of photoinduced charge transfer in donor-bridge-acceptor system.

4.2.3 Boundary of Superexchange and Hopping Model

For the case of long ranges, charge transfer could only be explained by a charge hopping model carried by the bridge base guanine or adenine radical cations. The majority of the experiments were concentrated on the rates of charge separation, shift, and charge recombination of DNA in solution using the multistep hopping model in long-range charge transfer studies.

On the other hand, in contrast to the hopping model, charge transfer over short-ranges was suggested to take place coherently via a superexchange mechanism rather than a hole hopping model.[208-230] A turning point at 3-4 bps from superexchange mechanism to hopping mechanism was established experimentally by Giese *et al.* on a well-defined system in which the charge transfer rate is investigated as a function of the number of A:T base pairs as bridges between donor and acceptor (see Figure 4.3 and Ref. [191]). They showed that the efficiency of hole transfer between a G and a GGG triplet strongly depends upon the length of the A:T bridge over short distance, but for a long length (n > 3), this efficiency changed very slightly with further increasing number of the A:T base pairs. A theoretical model for charge transport was created by Jakobsson *et al.*[231] using Monte Carlo simulation based on Marcus theory in order to properly describe the experimental observations. The theoretical calculations have shown a good agreement with the experimental results.

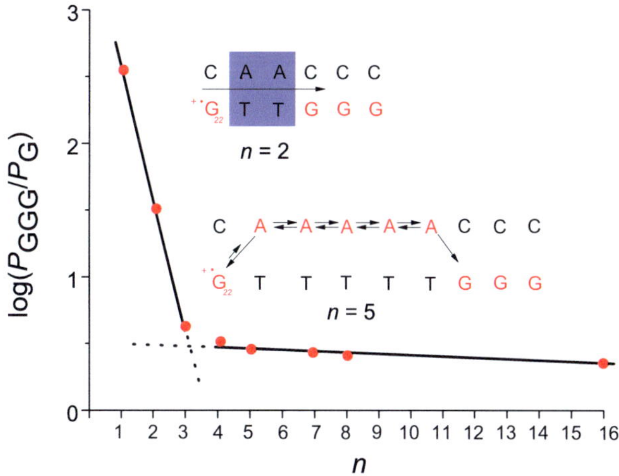

Figure 4.3: Plot of $\log(P_{GGG}/P_G)$ vs. the number (n) of the A:T base pairs. The turning point between superexchange and hopping model was obtained at $n = 2$. For superexchange regimes ($n < 3$), the rate constant of charge transfer is exponentially depending on the distances. For $n > 3$, an A-hopping mechanism takes place and shows a weak dependence on the distance. The first slope leads to $\beta = 0.6$ Å^{-1}.[191]

However, it is worth mentioning, that in many cases the transfer is expected to be governed by a mixture of the two mechanisms. Barton and coworkers proposed a generic mechanism of charge transfer as a mixture of superexchange and hopping mechanisms for a long-range charge transport in DNA.[232] The superexchange mechanism has been focused primarily on short-range charge transfer studies. In this case, however, the critical point to establish such a mechanism is that the energy gap of the bridge is much larger than the electronic coupling between donor and acceptor. In fact, one considers that for charge transfer over short distance regimes, these parameters are strongly dependent on the type of donor and acceptor applied.

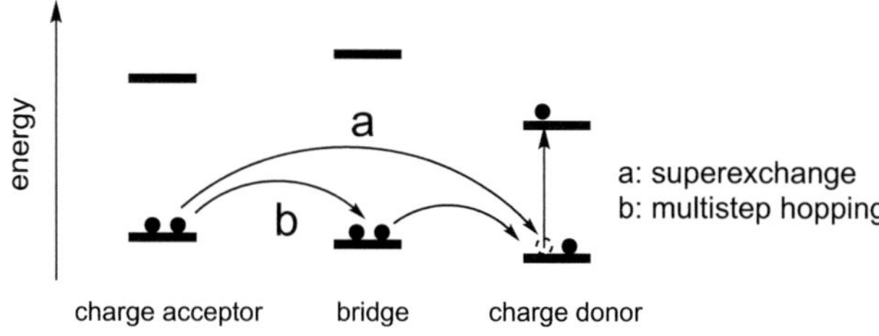

Figure 4.4: Simplified scheme for a combination between superexchange and hopping mechanisms which can occur in short-ranges charge transfer process.

4.3 Spectral Properties of the Donor-DNA-acceptor System

In order to understand the mechanism of charge transfer process through DNA over short-ranges, a new donor-DNA-acceptor system was synthesized for investigation (Figure 4.5) by the group of Prof. H.-A. Wagenknecht. In this system, Nile red (Nr) was used as a photosensitizer to inject a hole into DNA upon laser pulse excitation. It is noted that the band maximum of UV/Vis absorption spectrum is around 520 nm for Nr[233], however, the peak maximum is 100 nm red-shifted when Nile red is attached to the DNA strand. In order to better distinguish between them, the charge donor will be labeled as **Nr-dU** which is described as Nile red conjugated to the 5 position of 2'-deoxyuridine via an acetylene bridge (see Figure 4.5). **Nr-dU** shows an absorption peak at a wavelength around 620 nm, and it can be selectively excited with a 620 nm laser pulse since DNA shows an absorption band centered at about 260 nm. Photophysical properties of this system have been well characterized by Wagenknecht and coworkers.[234] 6-*N,N*-Dimethylaminopyrene is the charge acceptor and attached via its 2-position to the 5-position of 2'-deoxyuridine (**Ap-dU**).[235] The oxidation potential of **Ap-dU** was measured by cyclic voltammetry and has E_{ox} = 0.88 V (vs. NHE). Combined with the reduction potential of E_{red} = -0.75V (vs. NHE) and singlet-state energy of E_{00} = 2.1eV[236] for **Nr-dU**, the Rehm-Weller equation[237] (without Coulomb term): $\Delta G = E_{ox} - E_{red} - E_{00}$ gives a sufficient driving force of -0.50 eV for the photoinduced charge transfer process between **Nr-dU** and **Ap-dU**. In the above mentioned scenario (Section 4.2.3), the guanine and adenine radical cation must play the role of an intermediate charge carrier for the charge hopping process. The oxidetion potentials of guanosine and adenosine are E_{ox} = 1.29 V and E_{ox} = 1.42 V, respectively.[238] With regard to the estimated excited state potential of **Nr-dU** (E^*_{red} = 1.35 V), oxidation of guanosine is clearly feasible, but oxidation of adenosine seems to be a borderline case. Therefore, when **Nr-dU** was close to G-base, irradiation of the **Nr-dU** site with a 620 nm laser pulse can trigger the charge transfer between **Nr-dU** in the single excited state and adjacent G to give the **Nr-dU** radical anion (**Nr-dU**$^{\bullet-}$) and the G radical cation (G$^{\bullet+}$), respectively. Hence, both chromophore-modified DNA bases will be combined in double strands **DNA1-DNA6** (Figure 4.5) and these sequences will be varied in both the distance between donor and acceptor (from 1 base pair to 3 base pairs) and the sequence (C:G as intervening base pair for **DNA1-DNA3** or A:T for **DNA4-DNA6**). Control oligonucleotides (**Nr-dU-DNAn** without acceptor **Ap-dU**) were prepared as reference systems for comparison.

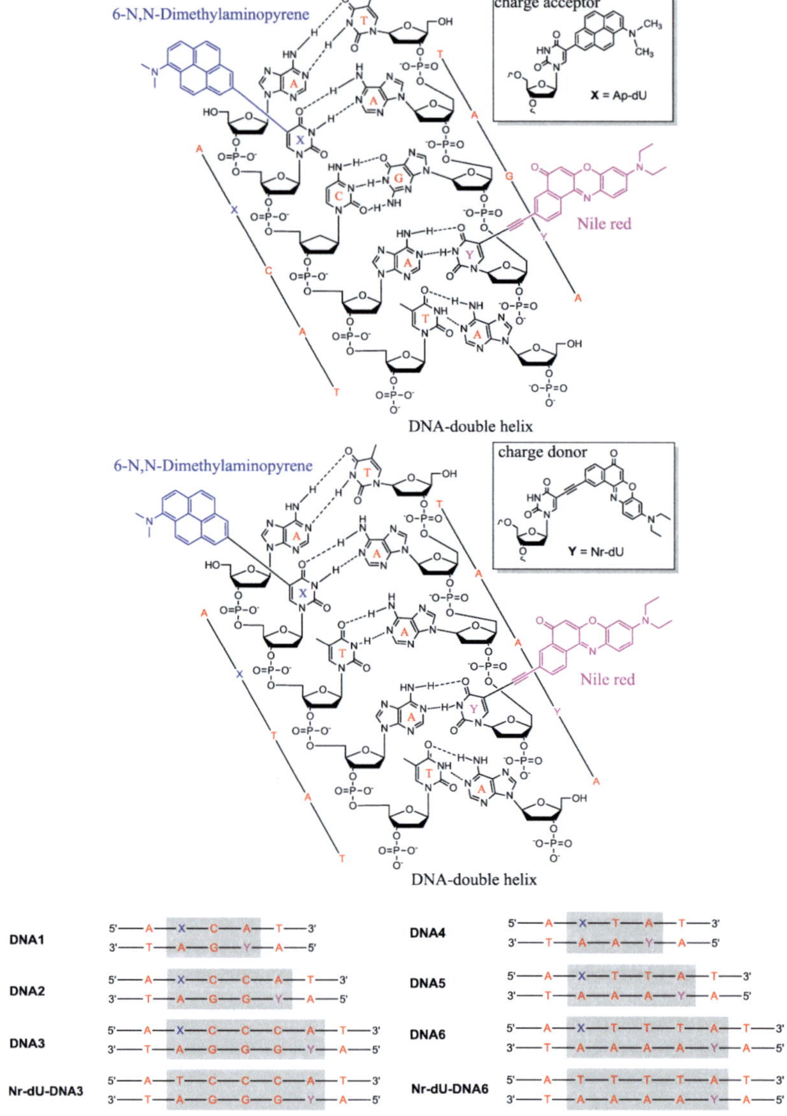

Figure 4.5: Structure of **Nr-dU** and **Ap-dU** and sequences of **DNA1-DNA6**.[210]

76

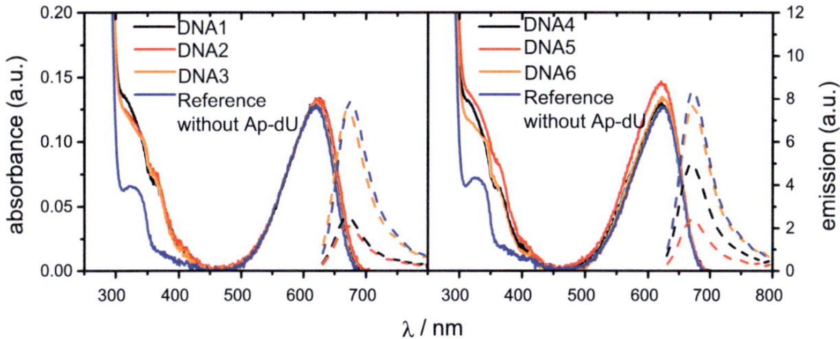

Figure 4.6: Absorption (solid) and emission (dash) spectra of **DNA1-DNA3** (left) and **DNA4-DNA6** (right) compared to the corresponding reference oligonucleotide duplexes (**Nr-dU-DNA3** for **DNA1-DNA3** and **Nr-dU-DNA6** for **DNA4-DNA6**; the spectra for **Nr-dU-DNA1**, **Nr-dU-DNA2**,**Nr-dU-DNA4** and **Nr-dU-DNA5** do not differ from the shown reference spectra); emission spectra were measured by excitation at 620 nm.[210]

Figure 4.6 shows the UV/Vis absorption and emission spectra of **DNA1-DNA6** and its corresponding references **Nr-dU-DNA3** and **Nr-dU-DNA6** for comparison. From the absorption spectra, the absorption band in the range from 300 to 400 nm exhibits the spectral overlap of the two chromophores in comparison with the references without acceptor **Ap-dU**.[210] The characteristic absorption band at 620 nm, however, which originates from the first single excited state of Nile red, can be used for selective excitation of **Nr-dU** in order to investigate the charge transfer process. The fluorescence measurements provided information on the overall quenching of the DNA series. Upon excitation at 620 nm the fluorescence intensity is observed decreasing as the distance between donor and acceptor is increased. Fluorescence quenching efficiencies for **DNA1-DNA6** are summarized in Table 4.1. This significant fluorescence quenching in the duplexes with one or two intervening base pairs (**DNA1**, **DNA2**, **DNA4**, and **DNA5**), but only minor quenching in **DNA3** and **DNA6** between **Nr-dU** and **Ap-dU** indicates a certain distance-dependence of the charge transfer process. It should be mentioned that the quenching efficiency is not increased linearly or exponentially by shortening the distance between donor and acceptor as expected. On the contrary, less efficiency quenching was observed in the shortest distance (68% for **DNA1** and 37% for **DNA4** compared with 73% for **DNA2** and **DNA5**, see Table 4.1).[210] This "anomalous" fluorescence quenching will be discussed together with the ultrafast time constants in the next section.

Another reason for the fluorescence quenching channel is electronic excitation energy transfer (EET) via the coupling of electronic excitations between energy donor and acceptor molecules. The rate of the energy transfer k_{EET} can be defined by[239,240]:

$$k_{EET} = \left(\frac{9000(\ln 10)}{128\pi^5 N_A n^4}\right)\frac{\kappa^2 Q_D}{\tau_D r^6} \cdot J(\bar{v}) \tag{4.3}$$

$$J(\bar{v}) = \int_0^\infty \frac{F_D(\bar{v})\varepsilon_A(\bar{v})}{\bar{v}^4}\,d\bar{v} \tag{4.4}$$

where $J(\bar{v})$ is the spectral overlap integral between the donor emission and the acceptor absorption. Obviously, Förster resonance energy transfer can take place only if the emission spectrum of the donor overlaps with the absorption spectrum of the acceptor and the both spectra should be located at separation distances within 1-10 nm from each other[241]. According to the absorption spectrum (central wavelength at ~350 nm) of **Ap-dU** and the emission spectrum (central wavelength at ~680 nm) of **Nr-dU**, fluorescence resonance, energy transfer between Nile red and the DNA-bridge, can be also excluded.

Moreover, spectroelectrochemical measurements (Figure 4.7) with the isolated nucleoside **Ap-dU** performed by the Wagenknecht group showed that the absorption bands near 290 and 363 nm collapse smoothly and two new spectral manifolds appear with their band maxima at 460 and 490 nm in similar intensity.[210]

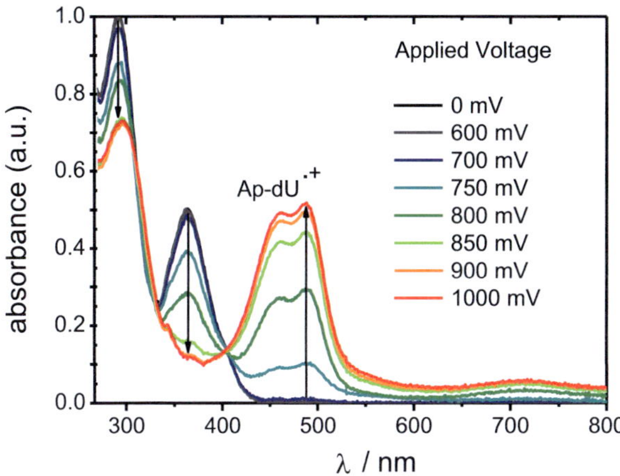

Figure 4.7: Spectroelectrochemical absorption measurements for **Ap-dU** in Dimethylformamide performed by the Wagenknecht Group.[210]

4.4 Charge Transfer through DNA

The main part of this section is to investigate charge transfer mechanisms and rates from the charge donor to the charge acceptor through DNA using femtosecond transient absorption spectroscopy. According to the UV/Vis absorption spectra of **Nr-dU**, the excitation wavelength was set at the absorption band maximum of 620 nm, and the probe wavelengths were varied over a broad spectral range (from 490 to 700 nm). The transient absorption of the photoexcited, doubly modified DNA samples consists mainly of two contributions: pump induced negative absorbance at ~620 nm and positive absorbance at ~490 nm.

For the first case, as expected, a strong negative absorption band by degenerate measurements shows that the transient profile of all samples is dominated by photoinduced bleaching of the first excited state of **Nr-dU** (Figure 4.8, left). In duplexes with a charge acceptor in the distance of 1 base pair (**DNA1** and **DNA4**) or 2 base pairs (**DNA2** and **DNA5**) the transient exhibits an additional rapid decay on a subpicosecond time scale in

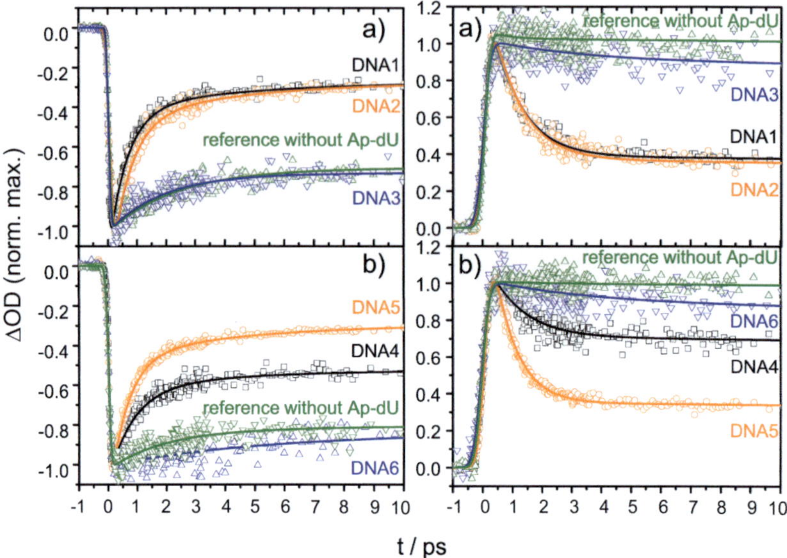

Figure 4.8: Experimental transient responses (symbol) and fitted curves (line) for a) **DNA1**, **DNA2**, **DNA3**, and reference without acceptor **Ap-dU**, b) **DNA4**, **DNA5**, **DNA6**, and corresponding reference. Left: Measurements by degenerated excitation and probing at 620 nm; Right: Measurements by excitation at 620 nm and probing at 490 nm.[210]

comparison with the corresponding reference. Distinguishingly, the profiles of **DNA3** and **DNA6,** intervening with 3 base pairs between donor and acceptor, present no subpicosecond temporal evolution after photoexcitation. The latter observation is in line with transients of duplexes in the absence of the charge acceptor, and resembles the ground-state bleaching of **Nr-dU**. These results suggest that over a "long" distance of 3 base pairs, an analogue subpicosecond temporal evolution is more difficult to be observed, and the transient resembles more that of the reference system. All negative transient profiles of **DNA1**, **DNA2**, **DNA4**, and **DNA5**, which can be fitted by a tri-exponential fit function, demonstrate similar behavior.

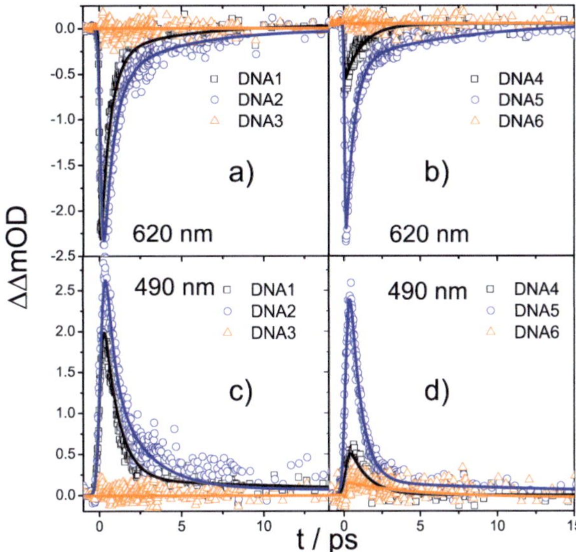

Figure 4.9: Retrieved transient responses of **DNA1-DNA6** after separation the contributions of corresponding reference systems. Please note that subtraction was accomplished with raw data sets and no scaling was necessary.[210]

Fitting of the profiles results in a very fast decay component and two slower components. The time constants are shown in Table 4.1. In contrast, a bi-exponential fit describes well the data of **DNA3**, **DNA6**, and the references with two slower time constants are comparable to the former systems, but with different relative amplitudes. In the second case, the origin of the positive absorbance at 490 nm seems to be more complicated (Figure 4.8, right). A fast subpicosecond time constant is observed in **DNA1**, **DNA2**, **DNA4**, and **DNA5**, while this component is absent in **DNA3** and **DNA6** as well as in the reference systems. On the one hand, this time constant may be partially attributed to the pyrene radical cation that is formed almost instantaneously (limited by the pulse duration) after photo-oxidation of **Ap-dU** by **Nr-dU** supported by spectroelectrochemical measurements. Combination of both measurements might give direct evidence for the presence of the $Ap^{\bullet+}$ radical cation. The corresponding subpicosecond component of the transient at 490 nm has obviously slower time decay, by 200 to 300 fs, compared with that at 620 nm (Table 4.1) and could therefore be assigned to a charge return process. On the other hand, in duplexes with long distances such as **DNA3**, **DNA6**, and the corresponding references, a strong positive transient absorbance appears likewise in the region of 490 nm, but the subpicosecond component is too weak to be observed. This observation, however, is important for the interpretation since obviously all transients in the vicinity of 490 nm are considered to be a superposition of radical absorption and higher excited state absorption (ESA) of the **Nr-dU**. ESA could mask the subpicosecond component in **DNA3** and **DNA6** in contrast to the one, or two base pairs intervening duplexes. This is better demonstrated if one subtracts transients from **DNA1-DNA6** by the corresponding reference system. These subtracted transients are shown in Figures 4.9 and 4.10. Here, early-time dynamics are only seen for **DNA1** and **DNA2,** as well as **DNA4** and **DNA5,** indicating radical dynamics (for time constants see Table 4.1).[210]

Sample	FQ (%)	Time constants for negative absorbance			Relative amplitudes for 620 nm		
		τ_1 (fs)	τ_2 (ps)	τ_3 (ps)	*rel. A₁*	*rel. A₂*	*rel. A₃*
DNA1	68	654 ± 47	2.7 ± 0.8	$\gg 100$	0.63	0.13	0.24
DNA2	73	582 ± 43	2.9 ± 1.2	$\gg 100$	0.62	0.15	0.23
DNA3	4	-	2.3 ± 0.4	$\gg 100$	-	0.31	0.69
DNA4	37	1038 ± 52	4.2 ± 0.8	$\gg 100$	0.39	0.10	0.51
DNA5	73	585 ± 46	4.2 ± 1.3	$\gg 100$	0.59	0.15	0.26
DNA6	8	-	4.4 ± 1.2	$\gg 100$	-	0.15	0.85

Table 4.1a: Time constants (τ_i) and relative amplitudes (A_i) resulting from global fits of the negative transients. (FQ: fluorescence quenching)[210].

Sample	Tm (°C)	Time constants for positive absorbance			Relative amplitudes for 490 nm		
		τ_1 (fs)	τ_2 (ps)	τ_3 (ps)	*rel. A_1*	*rel. A_2*	*rel. A_3*
DNA1	59.0	902 ± 51	-	$\gg 100$	0.81	-	0.19
DNA2	58.3	877 ± 53	-	$\gg 100$	0.82	-	0.18
DNA3	59.0	-	-	$\gg 100$	-	-	1.00
DNA4	58.6	1284 ± 64	-	$\gg 100$	0.49	-	0.51
DNA5	59.1	852 ± 44	-	$\gg 100$	0.80	-	0.20
DNA6	56.0	-	-	$\gg 100$	-	-	1.00

Table 4.1b: Time constants (τ_i) and relative amplitudes (*rel. A_i*) resulting from global fits of the positive transients. (Tm: melting temperature)[210].

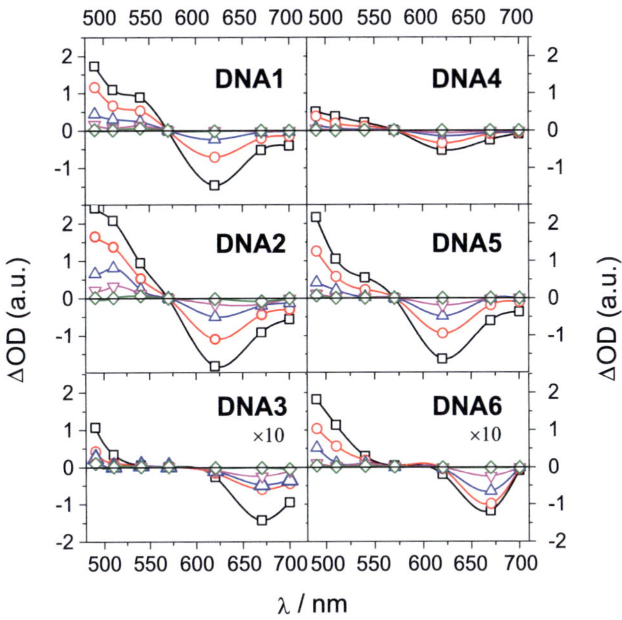

Figure 4.10: Subtracted transient absorption spectra of **DNA1-DNA6** that contain kinetics of charge transfer processes; 0.5 ps (black, square), 1 ps (red, circle), 2 ps (blue, up triangle), 5 ps (purple, down triangle), 10 ps (green, diamond).[210]

To begin a discussion on the time constants, one focuses at first on the first time constant (τ_1). Surprisingly, this time constant becomes larger for shortest distances (e.g., 654 fs for **DNA1** and 1038 fs for **DNA4**, in comparison with 582 fs for **DNA2** and 585 fs for **DNA5**). This phenomenon is similar to the observed fluorescence quenching efficiency as shown in Section 4.3 (68% for **DNA1** and 37% for **DNA4**, in comparison with 73% for **DNA2** and **DNA5**). The relative correlation between fluorescence quenching efficiency and this time constant was demonstrated in Figure 4.11. According to earlier studies of charge transfer over short-ranges, the transfer rates should be expected to be exponentially distance-dependent. In other words, the shorter the distance, the faster the charge transport, therefore the more efficient the fluorescence quenching. However, this ideal behavior is not observed in this donor-DNA-acceptor system. To explain this observation, two scenarios may be taken into account. Note that none of them can explain the dynamics alone. First, obviously, this time constant depends strongly on the distance between donor and acceptor. According to the experimental findings by Giese[191], the turning point between superexchange and hopping mechanism was obtained in **DNA3** as well as **DNA6**. A kinetics study by Schuster *et al.*[242] for hole transport through seven G:G pairs separated by $(A:T)_n$ (n = 2-5) bridges across the 3'-5' strand of the DNA duplex also reveals that the superexchange hole transfer crossover occurs at n = 3. For hole transfer via longer $(A:T)_n$ ($n \geq 3$) bridges, the superexchange mechanism is replaced by a parallel mechanism of thermally induced hole hopping via long $(A:T)_n$ chains. For charge transfer over short-ranges, the superexchange mechanism is characterized by an exponential D-A distance (R_{D-A}) dependence of the rate, allowing only for a short-range transfer, that is ≤ 10 Å[218,243] (in this case, $R_{D-A} \sim 6.8$ Å for **DNA1** or **DNA4**, and $R_{D-A} \sim 10.2$ Å for **DNA2** or **DNA5**). In general, the rate constants for nearest-neighbor hole transfer ($R_{D-A} \sim 3.4$ Å) with a low driving force are in the range of 10^{11}-10^{12} s^{-1}, and the rate constants for hole transfer through one intervening base pair (R_{D-A} ~ 6.8 Å)are observed between 10^{10}-10^{11} s^{-1}.[244] According to the experimental results, the first subpicosecond time constant at 600 fs (~1.67×10^{12} s^{-1}) could therefore be assigned to the rate of superexchange between donor and acceptor.

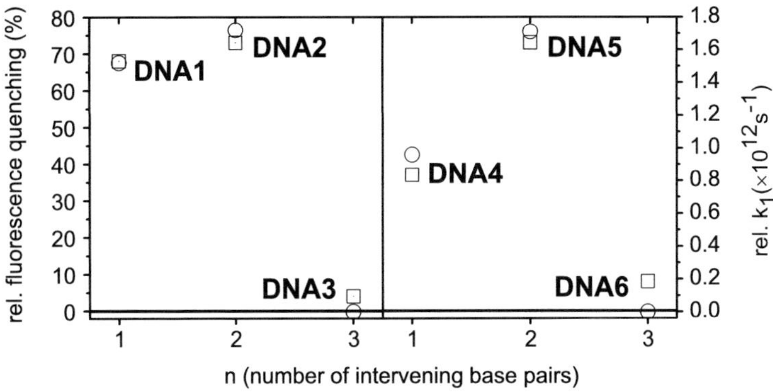

Figure 4.11: Relative correlation between the fluorescence quenching efficiency (square) and the first rate constant K_1 (circle) which obtained by transient measurements after degenerate excitation and probing at 620 nm. Plots were arbitrarily fixed for n = 1 for **DNA1**.[210]

Second, the "anomalous" behavior (slow time constant and less fluorescence quenching efficiency) of **DNA1** and **DNA4** may arise from structural hindrance (interaction between donor and acceptor) and reflect poor electronic coupling due to a structural distortion within the assembly. Although the B-form DNA is more stable in comparison with A- and Z-forms, the B-form structure can also be distorted in different ways, for example, by base-pair mismatches, the lack of a base in a base pair, or one or more "extra" molecules attached to the base site by synthesis, which result in the formation of a bulge in one strand of the duplex. Most studies with varied donor-DNA-acceptor systems assumed that the DNA duplexes still adopt an idealized B-form structure in order to simplify the investigation. Indeed,

this point cannot be neglected, since certain unusual behaviors were observed in some donor-DNA-acceptor systems, and thus trigger hot debate
about the transfer rates.[245] The same problem may exist in spectroscopic
investigation of the charge transfer using time-correlated single-photon
counting (TCSPC) and $1,N^6$-ethenoadenine as donor.[188] Although these
distortions may provide some uncertainty in determining charge transfer
rates, this issue is still important for the identification and interpretation of
fluorescence quenching and the influence of structural changes on the
charge transfer rate. According to the experimental findings, the rate constants correlate well with the measured fluorescence quenching efficiency
(Figure 4.11).

Sample	E_a(eV)	Time constants			Relative amplitudes		
		τ_1 (fs)	τ_2 (ps)	τ_3 (ps)	rel. A_1	rel. A_2	rel. A_3
DNA1	0.10	848 ± 55	4.9 ± 1.1	$\gg 100$	0.67	0.12	0.21
DNA2	0.07	793 ± 61	6.0 ± 0.8	$\gg 100$	0.63	0.17	0.20
DNA4	0.12	1147 ± 66	3.4 ± 1.9	$\gg 100$	0.45	0.10	0.45
DNA5	0.07	779 ± 32	4.3 ± 1.0	$\gg 100$	0.72	0.08	0.20

Table 4.2: Time constants (τ_i) and relative amplitudes (rel. A_i) resulting from fits of
the transients (1 °C) and the corresponding activation energy (E_a). [210].

Furthermore, if these subpicosecond time constants were ascribed to the
rate of superexchange mechanism only, then one would have to neglect the
temperature-dependent rates shown in Figure 4.12 and Table 4.2. Although
a weak temperature dependent rate was suggested for superexchange mechanism in molecular wires,[246] it is questionable whether this can be applied to the current system. Nevertheless, from the experimental results, the
temperature affects not only the first time constant, but also the second one
in the range of a few picoseconds. In addition, the retrieved activation energy matches the experimental as well as theoretical results of the tempera-

ture dependence for charge hopping rates between adjacent C:G base pairs, and can be well described by an Arrhenius-type law with an activation energy of 0.12 eV.[247,248] One may argue that the determination of an activation energy via only two different temperatures is not meaningful. However, physical properties limit the accessible temperature range of the current system significantly (between freezing, and melting point). Nevertheless, temperature dependence is observed for all samples allowing for an estimation of such a barrier. A possible explanation for this temperature dependence results from the loss of the driving force due to the competition between excited state relaxation of **Nr-dU** unit and charge transfer to the acceptor and reflects the change of reorganization energy in the charge transfer process by distance-dependent changes.[245b] From this point of

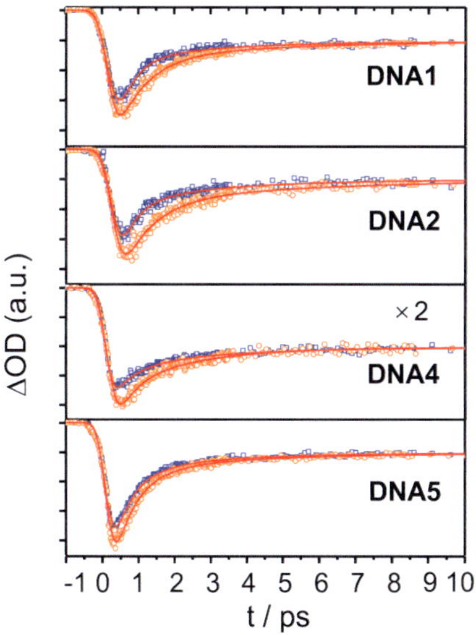

Figure 4.12: Temperature dependent transient responses (symbol) and fitted curves (line) for **DNA1**, **DNA2**, **DNA4**, and **DNA5** at room temperature (21 °C, blue square) and 1 °C (orange circle) after degenerate excitation and probing at 620 nm.[210]

view, one can raise the question whether superexchange and hole hopping mechanisms occur parallel over a short distance (Figure 4.4).

The attention here should be focused on the second and third time constants. As mentioned at the beginning, the ultrafast dynamics of the **Nr-dU** (ground state bleaching and excited state absorption) may mask the actual dynamics of charge transfer processes. This is better demonstrated if one subtracts the time profiles from **DNA1-DNA6** by its corresponding reference systems. Thus, the early-time constants may contain only the charge transfer kinetics. The longer time constant (>> 100 ps) can be assigned to the fluorescence lifetime of **Nr-dU**. This time constant is comparable to that of Nile red, although **Nr-dU** differs from Nile red in the UV/Vis absorption spectrum. Nile red has a fluorescence lifetime of about 4.5 ns.[249,250] From the UV/Vis absorption spectrum[250] one realizes that the band maximum around 520 nm for Nile red is 100 nm red-shifted when Nile red is attached to DNA (Figure 4.6). On the other hand, all long-time components can be fitted mono-exponentially such as for Nile red. In order to eliminate perturbation of Nile red with DNA-chains, one assumes that this long relaxation process (τ_3) exhibits the pure dynamics of **Nr-dU** fluorescence or ESA, the perturbation will be eliminated via subtraction of the long-time constant from transients of **DNA1-DNA6** with the corresponding reference samples (without acceptor **Ap-dU**). The residual transients presumably contain only two ultrafast time constants. Those time constants are similar to the original data which are shown in Table 4.1. The residual transients are presented in Figure 4.9 for the measurements by probe wavelength at 620 nm and 490 nm as an example.

Applying this separation method to all transients of probe wavelengths (from 490 to 700 nm), the new retrieved transient absorption spectra are obtained and shown in Figure 4.10. The general spectral features are similar to each other. A broad positive absorbance change is observed on the blue side below 570 nm. The first decay and the corresponding relative amplitude are still related to the fluorescence quenching efficiency of **DNA1-DNA6** not only for negative absorbance, but also for positive absorbance. This may suggest that the charge returns via the same channel. A negative absorbance around 620 nm (**DNA1**, **DNA2**, **DNA4**, and **DNA5**) exhibits charge transfer dynamics in comparison with the absence of a peak at 620 nm in **DNA3** and **DNA6**. The weak maxima at 670 nm in **DNA3** and **DNA6** can be interpreted as the residual fluorescence components, since the maxima emission of **Nr-dU** peaks at 670 nm (Figure 4.6).

As expected, the residual time constants on a subpicosecond timescale (~600 fs) are in good agreement with the original results, and correlate to the fluorescence quenching efficiency. This time constant can be ascribed to the rate of the superexchange mechanism. In contrast, the assignment of the second time constant (τ_2) on a time scale of a few ps with a relative weight of about 15% (see Table 4.1) seems to be more complicated. Obviously, this time constant is weakly distance-dependent, but clearly shows a base pair dependence (2.5 ps for C:G base pair and 4.5 ps for A:T base pair). It has been well established that the charge can transfer from the donor to the acceptor through the intermediate bridge via electronic superexchange interaction, if the energy of the bridge states are much higher compared to that of the donor and acceptor states. Therefore, theoretical calculations on the energy levels of donor, acceptor, and bridge states seem to be necessary. TDDFT-calculations were performed by Dr. T. Wolf in institute of physical chemistry (KIT) using the TURBOMOLE V-6.3 program package.[251] Ground state geometries were calculated with DFT-/BP86/def2-SV(P).[252-257] Singlepoint calculations were performed with DFT/B3LYP/aug-cc-pVDZ.[252-256,258] Singlet excitations were performed using TDDFT/B3LYP/aug-cc-pVDZ.[258-261]

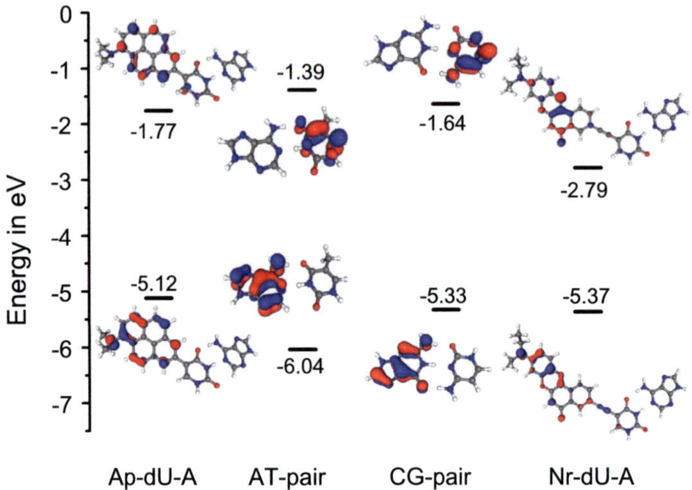

Figure 4.13: Theoretical calculations of the highest occupied molecular orbitals (HOMOs) and the lowest unoccupied molecular orbitals (LUMOs) of the base pairs A:T, C:G, **Nr-dU**-A, and **Ap-dU**-A.[210]

As shown in Figure 4.13, the HOMO and LUMO energies of A:T and C:G base pair, and the corresponding dye-substituted base pairs Nr-dU-A and Ap-dU-A, were calculated for better comparison. In the case of **Nr-dU-A**, singlet excitations were calculated to ensure that the excited transition corresponds to the HOMO-LUMO transition. The results reproduce nicely those published in the literature[262,263], that the HOMOs of the A:T and C:G base pairs are mainly located at the A and G base sites. Photo-excitation generates a hole in ground state (HOMO) of **Nr-dU-A** which is delocalized over the whole **Nr-dU** system. Inspection of the energy gaps and the location of the electronic density of the relevant MOs, for example, from the **Nr-dU-A** HOMO (-5.37 eV) to the C:G HOMO (-5.33 eV) located at G, suggests that the oxidation of the nearest-neighbor guanines by **Nr-dU*** seems to be very efficient in comparison with HOMO of A:T base pairs. As a consequence, based upon the calculated energy landscapes of donor (**Nr-dU**), acceptor (**Ap-dU**), as well as base pairs A:T and C:G, the bridge states are comparable in energy with the HOMO of the donor. Thus, a hole injection into DNA-bridge can take place on these bases. This suggests that, besides superexchange, hole hopping could be additionally taken into account in the systems investigated. Based on the determined time constants, the second time constant of a few ps can be assigned to the rate of hole hopping through the DNA-bridge. The results can be compared with estimations from other works on apparently different systems. For example, hole transfer by A-hopping has been studied by Majima and co-workers[264] using laser flash photolysis. They suggested that the hole transfer is weak distance-dependent over a distance range from 7-22 Å and the rate is higher than 10^8 s^{-1}. In another report, the transfer rate was estimated to be 2×10^{10} s^{-1} by analyzing the distance dependence of the quantum yields for charge separation in systems with a naphthaldiimide as acceptor and phenothiazine as donor. This method is regrettably limited by time resolution.[265] A weak distance-dependence has also been observed with ethidium as an oxidant for hole transfer reactions with 7-deazaguanine.[245a] In this report the rate constant was determined to be about 2×10^{11} s^{-1} for hole transfer, and the distance dependence between 10-17 Å, and suggested that the shallow distance-dependence for this reaction was attributed to transport within the hopping regime. On the other hand, photoexcitation of the donor can lead to thermal population of the bridge states. Another possible process for assignment of this time constant is a charge injected time scale before hole hopping through DNA duplexes helix. If this time constant is assigned to hole injection into DNA-bridge, it is then comparable to the results reported earlier for the system of ethidium and 7-deazaguanine, which gives a rate of (5 ps)$^{-1}$ for hole injection. Another proposed injection rate (10 ps)$^{-1}$ was reported for the adenine isomer 2-aminopurine system.[245c] For this reason, the second time constant of about 3 ps from this investigated system can also be assigned to the rate for hole injection.

4.5 Conclusion and Outlook

The time constants obtained in this short-range donor-DNA-acceptor system are in good agreement with the values in other charge transfer systems. An analysis of spectroscopic and theoretical calculated results suggested that charge transfer in this system may occur through both super-exchange and hopping mechanisms. Furthermore, the good correlation between fluorescence quenching efficiency and the first ultrafast time constant revealed one more and important insight for understanding the charge transfer mechanisms in DNA-mediated systems. However, as mentioned above, in order to explain the experimental observations, one assumes that the structural distortion in shortest distance system leads to anomalous behavior for fluorescence quenching efficiency and charge transfer rate. One needs to supply this assumption with an approximate theoretical calculation, which can be able to answer the remaining question about the influence of structural changes in DNA duplexes on the charge transfer rate constants. Furthermore, it remains an interesting and challenging aspect to study the charge transfer rates over long ranges in such donor and acceptor systems.

Chapter 5: Ultrafast Carrier Dynamics in Iron-lanthanide Clusters

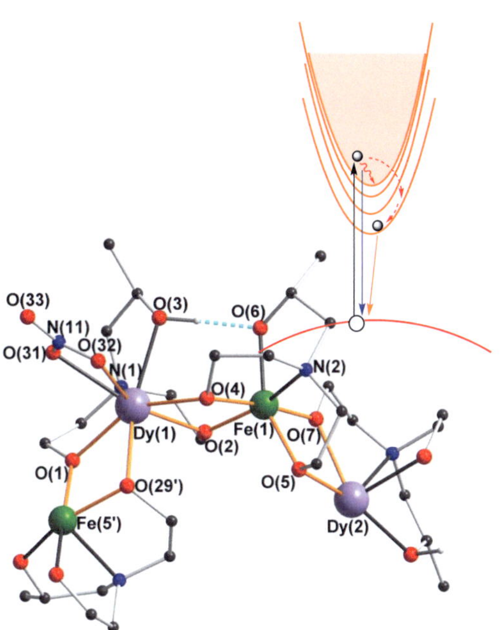

5.1 Introduction

From solid-state lasers to optical amplifiers in fiber optics, rare earth elements have been widely used in photonics materials.[266-269] In the last decade the chemistry of metal clusters have been extensively developed.[270] Most research focused on synthesis and investigation of magnetic properties. Using ions of lanthanides to modulate magnetic properties of heterometallic $3d/4f$ single-molecule magnets (SMMs) has become more common in recent years, mainly due to the magnetic anisotropy of some lanthanides which can increase the blocking temperature for magnetization reversal.[271,272] Conversely, from the photophysical point of view, inorganic nanocrystals doped with trivalent lanthanide (Ln^{3+}) ions with various behaviors have also triggered an increasing research topic. Lanthanide doped nanocrystals may differ significantly for the bulk affecting the luminescence properties and excited-state dynamics of the doped Ln^{3+} ions.[273] In particular, Luminescence of Ln^{3+} ions embedded in wide energy gap semiconductors can be sensitized by exciton recombination in the host materials, since direct excitation of the forbidden $4f\leftarrow 4f$ transitions is generally inefficient.[274] Those experimental findings have been applied in luminescent materials. The photophysical properties of lanthanide doped materials are important. In this project the investigation of such heterometallic Fe/$4f$ clusters will focus on ultrafast phenomena. The magnetic properties of these cyclic complexes were performed by the group of Prof. A. K. Powell using a combination of Mössbauer spectroscopy, electron paramagnetic resonance (EPR), and magnetic susceptibility measurements.[275]

The characteristic properties of iron-lanthanide clusters including samples synthesis, structure analysis, anisotropy, as well as magnetic properties have been represented in the dissertation of Dr. A. Baniodeh from Prof. Annie K. Powell's group (KIT). According to the results of this thesis, the magnetic properties of iron-lanthanide clusters (**Fe$^{(III)}_{10}$Ln$^{(III)}_{10}$:**

Ln = Pr, Nd, Sm, Eu, Gd, Tb, Dy, Ho, Er, Tm, Yb, Lu, Y) are different although they have the same structures. For example, the magnetic behaviors in most iron-lanthanide clusters exhibit ferromagnetic behaviors at low temperatures. However, in the case of the $Fe_{10}Tm_{10}$ cluster, the ferromagnetism becomes overwhelmed by antiferromagnetism.[275] The differences of the $Fe_{10}Ln_{10}$ clusters are shown not only in magnetic properties but also in color, ranging from colorless for $Fe_{10}Tm_{10}$ to red-orange for $Fe_{10}Er_{10}$, although the two elements (Tm, Er) are nearby in the periodic table.[275] Moreover, the main question during the synthesis is why the dark red color characterized for Fe^{3+} complexes arising from LMCT transitions disappears in $Fe_{10}Ln_{10}$ clusters. In order to get an insight into ultrafast dynamics (e.g., charge transfer between metal ions) within this assembly and to explain the quenching of LMCT states, the photophysical properties of those systems are investigated in this part. Using femtosecond laser pulses in a pump-probe technique, one can directly address the ultrafast processes which are responsible for the excitation and relaxation of an investigated system. In this section, the ultrafast dynamics of the iron-oxide system, namely carrier dynamics, will be presented using pump-probe measurements in combination with UV/Vis absorption spectroscopy. The influence of impurities (Ln^{3+}) on the ultrafast dynamics of iron oxide is demonstrated when compared with energy gaps of the corresponding iron-lanthanide cluster as determined by applying the Tauc equation.

5.2 Experimental Section

In the case of time-resolved pump-probe spectroscopy, the measurements were performed by means of another femtosecond laser system. The description of this femtosecond laser system has been provided elsewhere[276] and the schematic of its setup is illustrated in Figure 5.1. Briefly, the output of an amplified laser beam (CPA 2210, Clark-MXR) with a repetition rate of 1 kHz was split equally to pump two non-collinear optical parametric amplifiers (NOPAs, Clark-MXR). One of these two output pulses served for sum-frequency generation (310 nm with pulse duration about 100 fs (FWHM)) via a sum frequency mixing process with the center wavelength of the CPA (775 nm). The probe pulses were set between 500 and 600 nm by another NOPA and the energy of the pump pulses was varied from 0.4 to 1.0 μJ in order to check the excitation energy and probe wavelength dependence of the transients. Otherwise, experiments were performed at a pump energy of 0.8 μJ and probe wavelength at 550 nm. The probe energy was less than 0.1 μJ. The polarization of the probe pulses was set at magic angle (54.7°) by means of a tunable $\lambda/2$ plate (Alphalas). The measurements were carried out at room temperature.

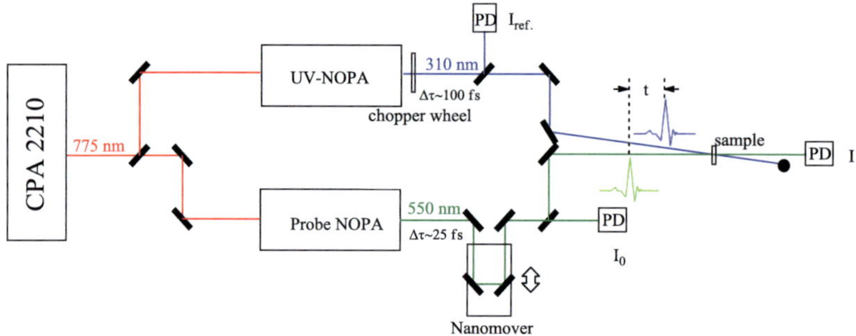

Figure 5.1: Schematic diagram of experimental setup used to measure the ultrafast dynamics of iron-lanthanide systems.

5.3 Structure of Iron-lanthanide Clusters

Most structures of **Fe$_{10}$Ln$_{10}$** clusters were determined at the ANKA Synchrotron, Karlsruhe.[275] The iron-lanthanide cluster, for example iron-dysprosium, shows a ring structure with alternating Fe^{3+} and Dy^{3+} ions (Figure 5.2a). The ring is built up from ten [FeDy(Me-tea)(Me-teaH)(NO$_3$)] repeating units (Figure 5.2c). Within one unit, Fe(1) is chelated by one nitrogen and three oxygen atoms of a fully-deprotonated (Me-tea)$^{3-}$

ligand, while Dy(1) is chelated by a doubly-deprotonated (Me-tea)$^{2-}$ ligand (Figure 5.2e). In both ligands, the deprotonated oxygen of their "ethanol" arms (here O(1), O(2), O(4) and O(5)) from *oxo*-bridge to adjacent metal centers in the ring. In contrast, the oxygen of the "*iso*-propanol" arms simply ligate their respective metal atoms without forming a bridge; the oxygens bonded to Dy^{3+} centers are still protonated (e.g., O(3)), while those bonded to Fe^{3+} centers are deprotonated (e.g., O(6)). Within a unit, one strong hydrogen bond with a distance between the two oxygens (O(3)$\cdots$O(6)) in the range of 2.52 -2.63 Å connected the two respective ligands (see Figure 5.2c. For more details about structure of **Fe$_{10}$Ln$_{10}$** clusters see Ref. [275]).

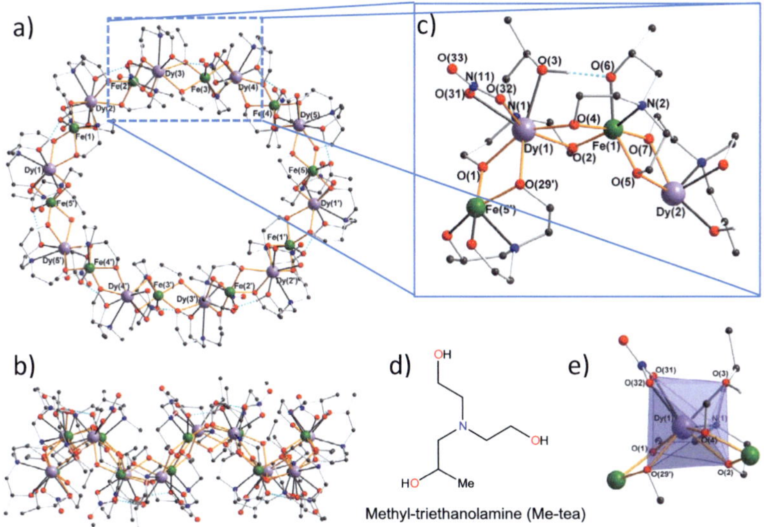

Figure 5.2: Molecular structure of the iron-dysprosium cluster: a) Top view and b) side view of the ring structure. c) The repeating unit of [FeDy(Me-tea)(Me-teaH)(NO$_3$)] is shown in the inset. Dysprosium in violet, iron in green, oxygen in red, nitrogen in blue, carbon in black, and hydro-bond shown as light blue dashed lines. d) Molecular structure of methyl-triethanolamine as ligand. e) Polyhedral coordinated dysprosium ions in **Fe$_{10}$Dy$_{10}$** cluster.[275]

5.4 Reference Systems

In order to better understand the ultrafast dynamics in **$Fe_{10}Ln_{10}$** systems, some reference systems are chosen for comparison. A good reference system having comparable ultrafast dynamics is one with a similar structural environment, like the systems investigated. In this case, obviously, the most ideal and favorable references are cyclic homometallic compounds containing only one kind of metal ions such as **$Fe^{III}_{10}Fe^{III}_{10}$** and **$Ln^{III}_{10}Ln^{III}_{10}$** clusters with the same coordination environment. Since **$Fe_{10}Fe_{10}$** clusters are not considered in this case, two different iron-oxide systems were checked for a selection as a reference. The structures of those iron-oxide clusters are shown in Figure 5.3. One of these two references (Figure 5.3a) is a ferric oxide compound [$Fe^{III}_7O_3(O_2CCMe_3)_9(teaH)_3$ $(H_2O)_3$] reported in Ref. [277] using Me-teaH$_3$ as a ligand (in the following denoted as **Fe_7-cluster**). Although this compound is not a ring structure like the one investigated in this case, it is still worth being employed for explanation of ultrafast dynamics. The second reference sample is also a cyclic ferric oxide cluster which contains six Fe^{3+} ions and six completely deprotonated $(tea)^{3-}$ ligands (Figure 5.3b, in the following denoted as **Fe_6-cluster**). This cluster has the coordination environment analogous to that of **$Fe_{10}Ln_{10}$** by using teaH$_3$ instead of Me-teaH$_3$ as ligand.[275]

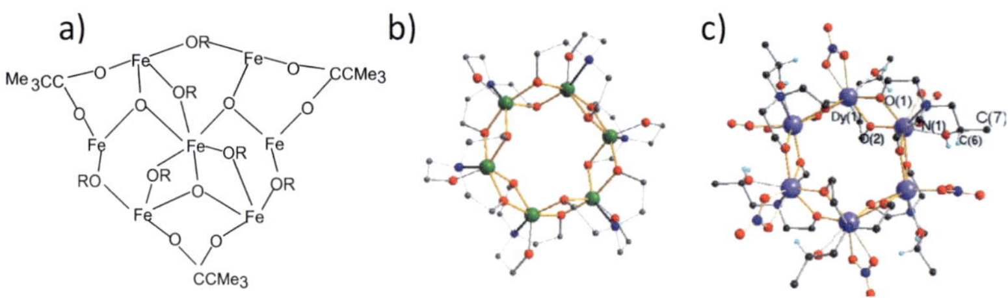

Figure 5.3: The structure of a) [$Fe^{III}_7O_3(O_2CCMe_3)_9(teaH)_3(H_2O)_3$][277], b) [$Fe^{III}_6(tea)_6$]·6MeOH as reference clusters. OR = Me-teaH$_3$ ligand [275], c) [Ln$_6$(Me-teaH)$_6$(NO$_3$)$_6$] (Ln = Tm and Dy).[275,278]

For $\mathbf{Ln_{10}Ln_{10}}$ clusters as reference, two clusters which contain only lanthanide ions were synthesized: $[Tm_6(Me\text{-}teaH)_6(NO_3)_6]\cdot6H_2O$ and $[Dy_6(Me\text{-}teaH)_6(NO_3)_6]$ (see Figure 5.3c, in the following denoted as $\mathbf{Tm_6}$- and $\mathbf{Dy_6}$-cluster). The synthetic process and magnetic properties were reported by Murray et al.[278] These homometallic clusters contain one unique Dy^{3+} ion coordinated with one $(Me\text{-}teaH)^{2-}$, one nitrate as well as one methanol solvent molecule. The repeating unit of Dy-oxygen exhibits an analog ring structure to $\mathbf{Fe_{10}Ln_{10}}$ clusters and can be used as reference compounds.

5.5 UV/Vis Absorption Spectroscopy

The UV/Vis absorption spectra of $\mathbf{Fe_{10}Ln_{10}}$ clusters in comparison with the reference $\mathbf{Fe_6}$-cluster are shown in Figure 5.4. All $\mathbf{Fe_{10}Ln_{10}}$ clusters including the $\mathbf{Fe_6}$ cluster have similar absorption spectra. A characteristic broad absorption band centers at about 200 nm with a shoulder at 300 nm, extending to the visible region. Since those clusters contain Fe-O-Ln repeating units, it is worth determining the origin of this transition at 300 nm. In other words, for pump-probe experiments, one should especially be interested in answering which part, or unit, in the cluster (e.g. Fe^{3+}, oxygen, Ln^{3+} or ligand) gives rise to this transition. In Figure 5.5, the UV/Vis absorption spectrum of $\mathbf{Fe_{10}Tm_{10}}$ clusters in the solid state[275] is shown. The corresponding spectrum for Fe_2O_3 nanoparticles is given in Ref. [279] for comparison. This bulk Fe_2O_3 showed a broad absorption band in the range of 200 – 500 nm with an absorption maximum at 300 nm. This high energy absorption band can be assigned mainly to the direct allowed ligand to metal charge transfer (LMCT) and partly to the Fe^{3+} ligand field transitions[279,280], which are characteristic of iron oxide nanomaterials.[279,281-283] The LMCT band has been associated with the electronic transition of O^{2-} to the t_{2g} and e_g orbitals of Fe^{3+} in the iron oxide cluster.[282-284] By the calculations of one-electron molecular orbitals of Fe^{3+} oxides and oxide hydroxides, the average energy of the quartet ligand fields states arising from the $(t_{2g})^3(e_g)^2$ configuration, have been estimated to be 325 nm.[282] However, three additional absorption bands peaking at 468, 690 and 795 nm are quite different from the iron oxide cluster (see Figure 5.5). These bands could be explained by doped Tm^{3+} ions in **$Fe_{10}Tm_{10}$** clusters. The studies of Tm^{3+}-doped glass[285-288] suggest that these bands arise from the $^3H_6 \leftarrow {}^1G_4$, $^3H_6 \leftarrow {}^3H_3$ and $^3H_6 \leftarrow {}^3H_4$ transitions of Tm^{3+} corresponding to 465, 685 and 790 nm, respectively. As a result, it can be determined that the absorption band at 310 nm in Figure 5.4 originates from a ligand-to-metal charge transfer (LMCT) absorption within the iron oxide unit in the **$Fe_{10}Ln_{10}$** clusters.

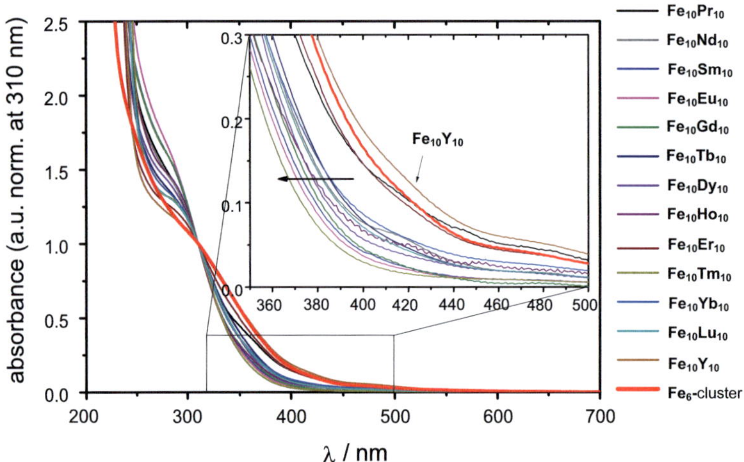

Figure 5.4: UV/Vis absorption spectra of **$Fe_{10}Ln_{10}$** and **Fe_6**-clusters in diluted solution. The spectrum of **Fe_6**-cluster is highlighted by red color.

The transitions originating from lanthanide ions in **$Fe_{10}Ln_{10}$** clusters can also be observed in solution. Figure 5.6a shows the UV/Vis absorption spectra of the solution of **$Fe_{10}Ho_{10}$** in comparison with that of $Ho(NO_3)_3$. The spectrum of Ho-nitrate shows several narrow and weak peaks between

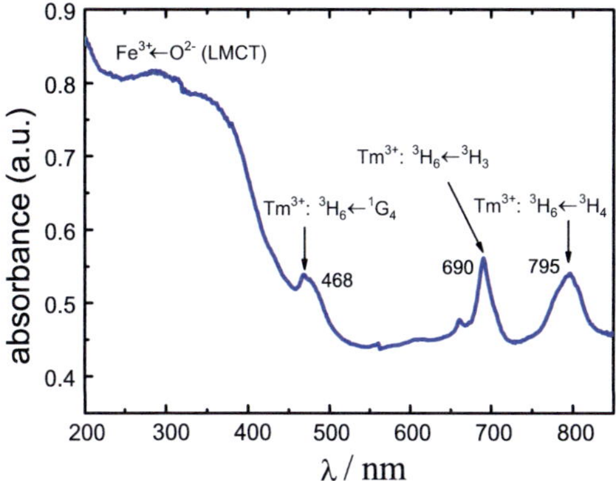

Figure 5.5: UV/Vis absorption spectrum of **Fe$_{10}$Tm$_{10}$** clusters in solid state.[275]

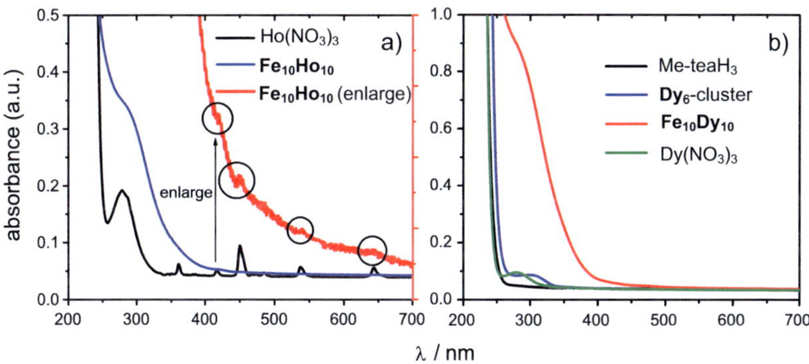

Figure 5.6: UV/Vis absorption spectra of a) **Fe$_{10}$Ho$_{10}$** (blue) clusters and Ho-nitrate (black) measured in a 1 mm cuvette. The curve in red shows an enlarged curve from **Fe$_{10}$Ho$_{10}$**; b) **Fe$_{10}$Dy$_{10}$**, **Dy$_6$**-clusters, Dy(NO$_3$)$_3$ as well as Me-teaH$_3$ ligand measured in a 1 mm cuvette.

350 and 700 nm which can be attributed to Ho^{3+} ions. It is well known that trivalent lanthanide ions possess unique spectral properties caused by the $4f$ - $4f$ electronic transitions within the $4f$ shell being shielded by the outer filled $5s^2$ and $5p^6$ shells. This shielding leads to weak interactions of the $4f$ electrons with their surroundings. Furthermore, there are only weak pertur-

bations of the electronic transitions between the energy levels of the $4f$ orbitals. As a consequence, this leads to nearly atom like narrow-line absorption and emission spectra with an excited state lifetime reaching the millisecond timescale.[289-292] Meanwhile, in the spectrum of **Fe$_{10}$Ho$_{10}$**, those peaks arising from Ho^{3+} ions can be observed when one enlarges the spectrum (Figure 5.6a, red). This indicates that the broad band observed at 300 nm cannot be assigned to any transitions within lanthanide ions. Further UV/Vis absorption spectroscopic measurements were performed in **Fe$_{10}$Dy$_{10}$**, **Dy$_6$**-cluster, Dy-nitrate as well as the Me-teaH$_3$ and teaH$_3$ ligands for comparison. The spectra are shown in Figure 5.6b. The lanthanide absorption bands are not observed in the absorption spectrum, because of their low absorption coefficients. The absence of the strong absorption at 300 nm in the **Dy$_6$**-cluster, Dy-nitrate and the corresponding ligands indicates that those compounds should not be taken into account in further pump-probe experiments. Moreover, UV/Vis absorption spectra of γ-Fe$_2$O$_3$ and α-Fe$_2$O$_3$ nanoparticles have been reported and exhibited a broad absorption from the visible to UV region, with a maximum at ~300 nm.[279,293] As a conclusion, according to the experiments and literature, the peak at 300 nm is assigned to LMCT within the iron oxide unit in **Fe$_{10}$Ln$_{10}$** clusters. Therefore, pump-probe experiments, in which the excitation wavelength was set at 310 nm, were focused on the ultrafast dynamics in iron oxide unit. The ultrafast dynamics will be represented in the next section.

5.6 Ultrafast Dynamics in Iron-lanthanide Clusters

All transients of **Fe$_{10}$Ln$_{10}$** clusters as well as the **Fe$_6$**-cluster are shown in Figure 5.7 and can be fitted with tri-exponential functions delivering three time constants: the first time constant (τ_1) on subpicosecond timescale and the second one (τ_2) within a few picoseconds, the third time constant (τ_3) on the order of about 100 ps. All time constants and corresponding relative

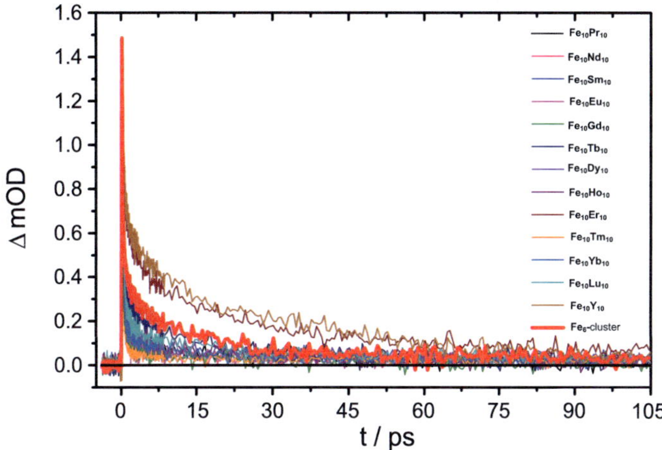

Figure 5.7: Temporal profiles of **Fe₁₀Ln₁₀** and **Fe₆**-clusters in diluted solution after excitation at 310 nm and probing at 550 nm. The spectrum of **Fe₆**-cluster is highlighted by red color.

amplitudes defined by $rel.A_i = A_i/\sum A_i$ are summarized in Table 5.1. Reference measurements in $Ho(NO_3)_3$, $Dy(NO_3)_3$, **Dy₆**-cluster, **Tm₆**-cluster, as well as pure ligand in solvent gave no a significant signals under the same experimental conditions. In contrast, the transients of **Fe₆** and **Fe₇**-cluster show a similar behavior in comparison to that of **Fe₁₀Ln₁₀**.

Before discussing these three ultrafast time constants of **Fe₁₀Ln₁₀** clusters, it is worth comparing the experimental results of the reference systems with literature values from earlier studies which might provide useful information about the ultrafast dynamics. Since references **Fe₆** and **Fe₇**-cluster contain repeating Fe-O units, the ultrafast dynamics of Fe_2O_3 materials therefore can be compatibly used to explain the dynamics of these reference systems. The ultrafast dynamics for the Fe_2O_3 nanoparticles was studied by Cherepy et al.[279] using pump-probe techniques. Three different time constants 0.36, 4.2 and 67 ps (see Ref. [279] and Table 5.1) are reported. Colloidal nanocrystal heterodimers composed of a gold domain and an iron oxide domain have been investigated by means of femtosecond transient absorption spectroscopy in order to obtain the charge transfer at the interface between gold and iron oxide.[294] In this investigation, the time scales of ultrafast dynamics in iron oxide single nanocrystals were found to be: $\tau_1 = 120 \pm 50$ fs, $\tau_2 = 4 \pm 1$ ps, and $\tau_3 > 100$ ps. By the study of carrier dynamics in α-Fe_2O_3 (0001) thin films and single crystals by femtosecond transient absorption experiments, it was obtained that the relaxation of the "hot" electron to the band edge was within 300 fs, the recombination or trapping of holes was within 5 ps, and the lifetime of trapped states was within hundreds of picoseconds.[295] Moreover, femtosecond laser spectroscopic experiments of a Fe^{3+}-Nafion membrane showed a relaxation dynamics for the excited state close to that observed in α-Fe_2O_3 nanocrystallite colloidal solution as a reference, which resulted in three time constants: $\tau_1 = 142 \pm 30$ fs, $\tau_2 = 4.5 \pm 0.9$ ps, $\tau_3 = 30$ ps and the first relative amplitude *rel.* $A_1 = 0.65$. In addition, the tri-exponential transient absorption decay of α-Fe_2O_3 in SO_3^--water clusters was also obtained with time constants being close to 320 fs, 1.5 ps and 31 ps.[296] Consequently, our data agree well with that obtained by previous studies. This means that the dynamics is dominated by the iron oxide domain in our systems, which also has been discussed in Section 5.3.3. Thus, a better understanding of the dynamics in Fe_2O_3 materials is the key to understand the dynamics in the systems investigated.

Clusters	Time constants			Relative amplitudes		
	τ_1 (fs)	τ_2 (ps)	τ_3 (ps)	*rel.* A_1	*rel.* A_2	*rel.* A_3
$Fe_{10}Pr_{10}$	191 ± 20	2.1 ± 0.1	45 ± 4	0.81	0.13	0.06
$Fe_{10}Nd_{10}$	272 ± 16	2.5 ± 0.2	99 ± 20	0.75	0.21	0.04
$Fe_{10}Sm_{10}$	230 ± 15	2.7 ± 0.4	131 ± 20	0.82	0.11	0.07
$Fe_{10}Eu_{10}$	298 ± 31	2.9 ± 0.4	121 ± 19	0.88	0.07	0.05
$Fe_{10}Gd_{10}$	197 ± 18	1.9 ± 0.2	39 ± 6	0.79	0.15	0.06
$Fe_{10}Tb_{10}$	258 ± 22	4.8 ± 0.6	45 ± 5	0.75	0.17	0.08
$Fe_{10}Dy_{10}$	219 ± 19	2.7 ± 0.3	102 ± 13	0.77	0.14	0.09

Fe$_{10}$Ho$_{10}$	251 ± 26	4.4 ± 0.6	84 ± 12	0.78	0.12	0.10
Fe$_{10}$Er$_{10}$	196 ± 18	3.2 ± 0.6	58 ± 10	0.64	0.20	0.16
Fe$_{10}$Tm$_{10}$	305 ± 30	4.2 ± 0.7	150 ± 22	0.88	0.07	0.05
Fe$_{10}$Yb$_{10}$	198 ± 21	2.5 ± 0.2	74 ± 12	0.72	0.16	0.12
Fe$_{10}$Lu$_{10}$	293 ± 24	2.6 ± 0.3	67 ± 6	0.69	0.17	0.14
Fe$_{10}$Y$_{10}$	382 ± 27	6.7 ± 1.2	40 ± 14	0.52	0.24	0.24
Fe$_6$	190 ± 15	5.2 ± 1.4	38 ± 11	0.61	0.23	0.16
Fe$_7$	367 ± 16	4.8 ± 0.8	57 ± 8	0.61	0.22	0.17
Fe$_2$O$_3$[279]	360 ± 84	4.2 ± 1.6	67 ± 36	0.66	0.22	0.12

Table 5.1: Time constants (τ_i) and corresponding relative amplitudes (*rel. A_i*) of **Fe$_{10}$Ln$_{10}$** clusters and reference **Fe$_6$**- and **Fe$_7$**-cluster in comparison to Fe$_2$O$_3$-nano-particles[279].

These earlier studies of Fe$_2$O$_3$ materials mentioned above suggested that photoexcitation gives rise to direct transitions of electrons from the *2p*-valence band orbitals of O^{2-} to the conduction band. The fast decay time of excited electrons is within the range of a few picoseconds while the trapping into trap states under the conduction band edge (with subpico-second timescales) is the predominant phenomenon.[279,294,295] In single-crystalline Fe$_2$O$_3$ hollow nanocrystals, Fan *et al.*[297] observed an ultrafast relaxation and also subsequent slower relaxations upon excitation at 400 nm originating mainly from the LMCT. They suggested that the ultrafast decay within 100-200 fs can be assigned to an initial fast relaxation of hot electrons to the conduction band edge. The second decay occurring on several picoseconds is attributed to the relaxation of the hot electrons to trap states under the conduction band or recombination of hot electrons to the valence band. The slow decay within tens to hundreds of picoseconds time scales exhibited the lifetime of trapping electrons.[297] The same ultrafast relaxation dynamics were also suggested by Joly *et al.* for carrier dynamics in α-Fe$_2$O$_3$ (001) thin films.[295] As a consequence, the three time constants observed in Fe_2O_3 materials can be well explained using band structure theory in semiconductors. Since $\mathbf{Fe_{10}Ln_{10}}$ clusters can be considered as lanthanide ion doped iron oxide clusters with high impurities, the lanthanide ions create a bunch of trap states under the edge of the energy gap in the clusters. This level acts as a trap to bind either hot electrons or holes. However, the size of single $\mathbf{Fe_{10}Ln_{10}}$ cluster is about 2-3 nm, much smaller than that of the widely studied Fe_2O_3 materials which are usually larger than 10 nm.[275] Thus, in these iron-lanthanide clusters, it is difficult to say whether the hot electrons delocalize in a conduction band around the ring, or localize in an LMCT transition state inside the iron oxide unit. One might assume that there is a quasi-continuous conduction band according to the higher lying LMCT transitions above 3 eV in giant molecular materials (see Ref. [282] and Section 5.3.3). Nevertheless, the time constants and corresponding relative amplitudes measured in $\mathbf{Fe_{10}Ln_{10}}$, as well as $\mathbf{Fe_6}$- and $\mathbf{Fe_7}$-reference systems, are in good agreement with the literature values indicating analogue relaxation processes. However, there is no report on the effects of doping of Fe_2O_3 nanoparticles by trivalent ions such as Ln^{3+} and ultrafast relaxation processes. Coming back to the investigated systems, as sketched in Figure 5.8, after photoexcitation an electron undergoes a transition from the valence band to the conduction band. The electron in the conduction band is called the hot electron. The excitation wavelength corresponds to the distance between valence and conduction band, namely the energy gap, while the rest left in the valence band is called a hole. This hole can be treated as a positively charged particle. However, the electron in the conduction band and the hole in the valence band do not separate immediately, but form a tightly-bound charge pair via Coulomb interaction, which is usually defined as localized exciton.[298,299] Under the conduction band edge there are several trap states, which can be caused by defect of the cluster, the coordinated ligand, the surrounding or impurities doped in the cluster.[300] In the investigated systems, most of these states might arise

106

from the doped lanthanide ions. The hot electron relaxes to the minimum of the conduction band and/or traps rapidly into the trap states within 200-400 fs. In general, an increasing impurity concentration in the semiconductor might give rise to a high density of trap states and therefore fast and efficient trapping of hot electron into these trap states. In comparison with the first time constant (190 ± 15 fs), the slower τ_1 for $\mathbf{Fe_{10}Ln_{10}}$ clusters is possibly due to competing relaxation channels including vibronic relaxation to the band edge and trapping to the trap states. The second time constant (τ_2) of a few picoseconds can be assigned to the recombination of the hot electron to the hole, releasing energy to the surrounding instead of emitting photons. The slowest time constant (τ_3) exhibits the lifetime of the trap states. Although the ultrafast processes dominate within the iron oxygen domain, these three corresponding time constants are essentially different with various doped lanthanide ions (Table 5.1). These important differences may result from the influence of the lanthanide ions.

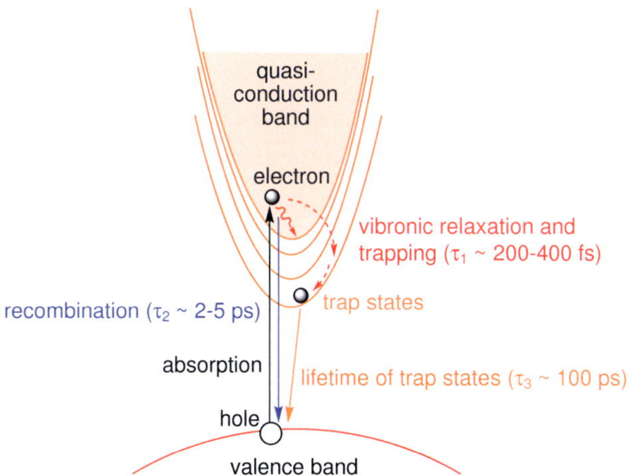

Figure 5.8: Scheme for ultrafast carrier dynamics in a ligand-metal charge transfer (LMCT) state in an iron oxide domain.

Now the influence of lanthanide ions on the ultrafast dynamics will be discussed. A simplified summarization of the first and second time constant correlated to corresponding relative amplitudes is plotted in Figure 5.9. In general, the influence of lanthanide ions on the first time constant (τ_1) leads to a slow decay but to a higher efficiency (high relative amplitude *rel. A_1*) compared to the **Fe$_6$**-cluster. In contrast, the effect of Ln^{3+} gives rise to a quick recombination (τ_2) of the hot electron back to the hole. However, there is not a trend in all time constants and relative amplitudes through the trivalent lanthanide ion series. Significant differences with opposed behaviors were observed in **Fe$_{10}$Er$_{10}$**, and **Fe$_{10}$Tm$_{10}$** clusters, which also show differences in color and magnetism. The first time constant and the corresponding relative amplitude (0.64) of **Fe$_{10}$Er$_{10}$** is similar to that of the **Fe$_6$**-cluster (see Table 5.1), and therefore they have the analogue physical behaviors (in color and magnetism). In contrast to **Fe$_{10}$Er$_{10}$**, **Fe$_{10}$Tm$_{10}$** and **Fe$_{10}$Eu$_{10}$** have the highest first relative amplitudes (0.88), indicating that the trapping of hot electron into trap states is more efficient. The influence of other lanthanide ions on the dynamics seems to be more complicated and do not follow any obvious rules. Nevertheless, although it is known that the 4*f*-electrons of lanthanide ions have weak effects on the surrounding due to the shielding effect by 5*s*- and 5*p*-electrons, experimental results clearly show significant effects on the ultrafast dynamics. It is worth mentioning here that the significant effects on the surrounding between thulium and erbium ions are not observed for the first time. Earlier studies on the electron-phonon coupling strength of nine lanthanide ions in $LiYF_4$-host lattices by Elles *et al.*[301,302] using temperature-dependent line broadening measurements reported that Tm^{3+} shows an efficient interaction with phonons from lattices in contrast to Er^{3+}. For the whole series of lanthanide ions in the periodic table, the interaction shows a symmetric behavior: the electron-phonon coupling is strong in the beginning, and at the end of the series, while it is weak in the middle. However, until now, this has not led to a further understanding of a possible variation of the electron-phonon coupling strength through the trivalent lanthanide ion series.[302] In the case of $Fe_{10}Y_{10}$, the effect on the ultrafast dynamics behaves exactly opposite to that of $Fe_{10}Ln_{10}$, as if it does not belong to one of the lanthanide elements. It is well known that, under the rare earth element term one usually includes also scandium (Sc) and yttrium (Y) although they have a different electronic configuration (without $4f$ electrons).[303,304] However, with the absence of electrons in the $4d$ and $4f$ electron shells the Y^{3+} ion is in a strict sense a transition metal ion rather than a lanthanide ion.

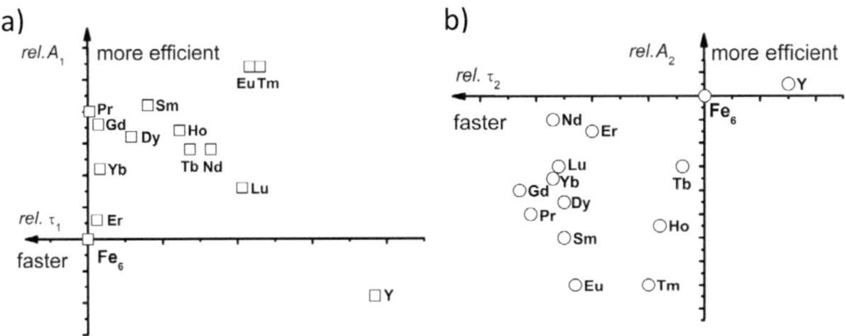

Figure 5.9: Plot of the relaxation decay τ_i versus corresponding relative amplitude *rel. A*$_i$ ($i = 1$ for a) and 2 for b)). The Fe_6-cluster is set to the zero reference point.

Note that the excitonic relaxation dynamics might cause exciton-exciton annihilation, when two or more excitons are generated in one Fe-O unit. Further experiments were performed using pump-intensity dependent pump-probe technique in $Fe_{10}Er_{10}$ clusters, for example. As shown in Figure 5.10, the differences between high- and low-intensity excitation in the decay profiles are not significant. Their corresponding maximums of amplitudes, and the amplitudes at 5 ps of the transient absorption responses, exhibit a linear dependence on the excitation energy. This suggests the possibility of localized excitation, and therefore, it can be attributed to exciton annihilation processes (one exciton pro Fe-O unit). Moreover, probe wavelength-dependent studies showed a wavelength-independent behavior (see Appendix C.1). Those results are in good agreement with that in Fe_2O_3 nanoparticles.[279]

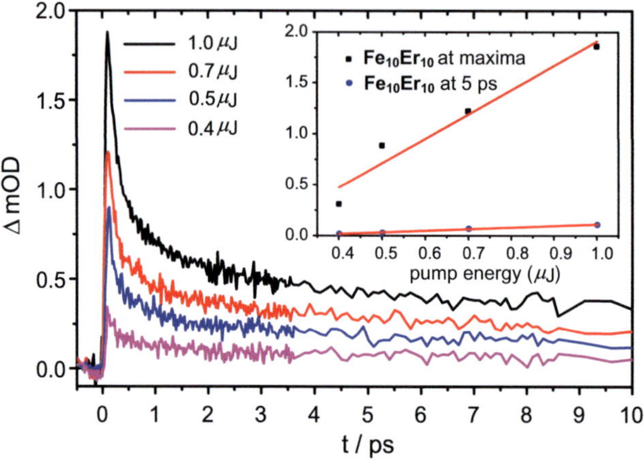

Figure 5.10: Transient absorption responses of **$Fe_{10}Er_{10}$** with different pump pulses intensities (1.0, 0.7, 0.5 and 0.4 μJ) and probe at 550 nm. The inset shows the corresponding amplitudes of the transient absorption scale linearly with the pump intensity.

5.7 Energy Gaps

The UV/Vis absorption spectra of **$Fe_{10}Ln_{10}$** clusters do not differ much from the spectrum of the **Fe_6**-cluster as well as Fe_2O_3 nanoparticles which are usually called bulk material.[305] As shown in Figure 5.4, the spectra exhibit a sharp step-function-like increase in absorption at energies close to the energy gap, which generally refers to the energy difference between the conduction and valence band. The optical energy gap with allowed transitions can be determined using the Tauc's equation:[306,307]

$$\alpha h v = A_0 \left(h v - E_{opt} \right)^n \tag{5.1}$$

where α is the absorption coefficient measured as a function of photon energy hv, and A_0 is a constant determined by the transition probability and related to the properties of the materials. E_{opt} is the optical energy gap and n is a number that depends on the type of transition process where $n = 1/2$, 2, 3/2, or 3 corresponding to allowed direct, allowed indirect, forbidden direct, and forbidden indirect electronic transition, respectively. Since the transitions in Fe_2O_3 are allowed direct, the Tauc equation requires $n = 1/2$ for all iron-lanthanide clusters. Therefore, a straight line fit to a plot of $(\alpha h v)^2$ versus hv gives a linear version of the Tauc equation as shown in Figure 5.11 with the reference **Fe₆**-cluster as an example. Here the optical energy gap E_{opt} value is determined by the intersection of the curve tangent with the x-axis. Its value is found to be 3.48 eV. This optical energy gap is much larger than in other α-Fe_2O_3 nanoparticles (~2.13 eV)[308] due to the ring structure. Since semiconductors have typical energy gaps of one or two eV or less, while the energy gaps of typical insulators are of order 5-10 eV,[309] the iron lanthanide clusters can be considered as a quasi-semiconductor with energy gaps around 3.5 eV. It is worth mentioning that the size of bulk matter can influence the electron energy gaps.[310] In bulk matter, as the size increases and reaches the nanometer scale, the overlap of numerous orbitals leads to a decrease of the energy levels and therefore narrow the width of the energy gap.[311,312] It can be observed that the energy gap displays a blue shift with decreasing size.[313-315] Thus, the ideal reference **Fe₁₀Fe₁₀** may have a different energy gap in comparison with the **Fe₆**-clusters. The energy gap values of **Fe₁₀Ln₁₀** can be calculated using Tauc equation and the results are shown in Figure 5.12 and summarized in Table 5.2.

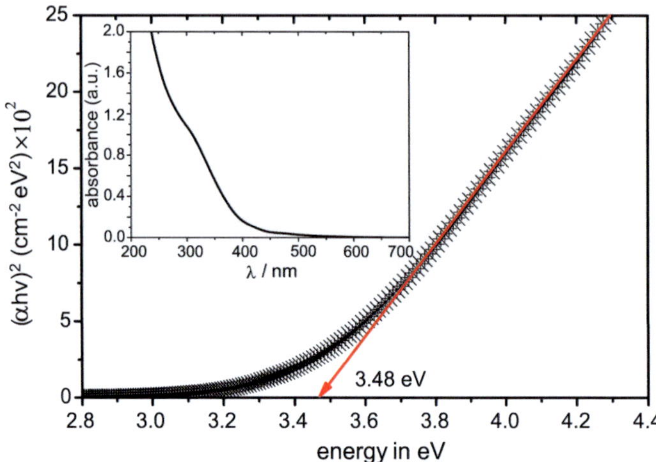

Figure 5.11: The Tauc plot shows the relationship between the incident photon energy (*hv*) and the absorption coefficient (α) near the absorption edge of **Fe₆**-cluster. The optical energy gap (E_{opt}) can be calculated according to 3.48 eV. Inset shows the absorption spectrum of the **Fe₆**-cluster.

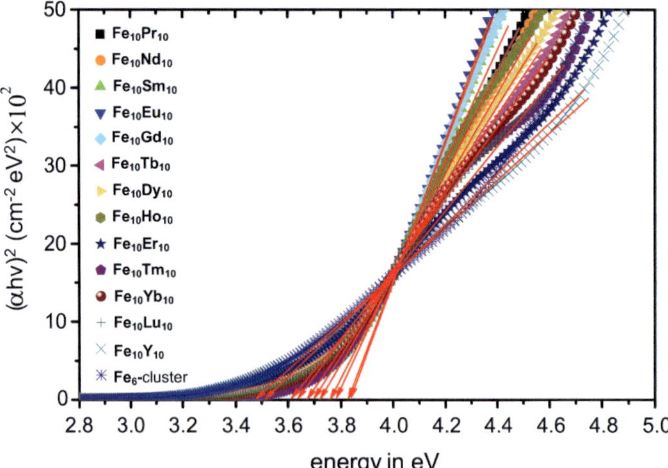

Figure 5.12: Tauc plot showing the position of the optical energy gap values for **Fe₁₀Ln₁₀** clusters.

Cluster	Energy gap (eV)	Cluster	Energy gap (eV)	Cluster	Energy gap (eV)
$Fe_{10}Pr_{10}$	3.76	$Fe_{10}Tb_{10}$	3.67	$Fe_{10}Yb_{10}$	3.69
$Fe_{10}Nd_{10}$	3.73	$Fe_{10}Dy_{10}$	3.71	$Fe_{10}Lu_{10}$	3.67
$Fe_{10}Sm_{10}$	3.80	$Fe_{10}Ho_{10}$	3.77	$Fe_{10}Y_{10}$	3.51
$Fe_{10}Eu_{10}$	3.85	$Fe_{10}Er_{10}$	3.61	Fe_6	3.48
$Fe_{10}Gd_{10}$	3.79	$Fe_{10}Tm_{10}$	3.73		

Table 5.2: The optical energy gaps (E_{opt}) of $Fe_{10}Ln_{10}$ clusters and reference sample Fe_6-cluster determined using Tauc's equation.

Obviously, the energy gaps of $Fe_{10}Ln_{10}$ are larger than that of the reference Fe_6-cluster. The influence of lanthanide ions on the ultrafast dynamics in $Fe_{10}Ln_{10}$ clusters by pump-probe studies is comparable to that on the energy gaps. A direct comparison between the energy gaps with the first relative amplitudes of the first relaxation process is shown in Figure 5.13. It is clear that the energy gaps are well correlated with the first relative amplitudes. If the first relative amplitudes were assigned to the efficiency of trapping process, then the larger the energy gap is, the more efficient hot electrons trap. From the UV/Vis absorption spectra of $Fe_{10}Ln_{10}$ clusters, addition of Ln^{3+} ions led to an increase in the effective energy gap and consequently the blue shift in the absorption threshold (Figure 5.4 inset). This upward band increasing suppresses, therefore, the electron-hole recombination processes leading to more efficient trapping.

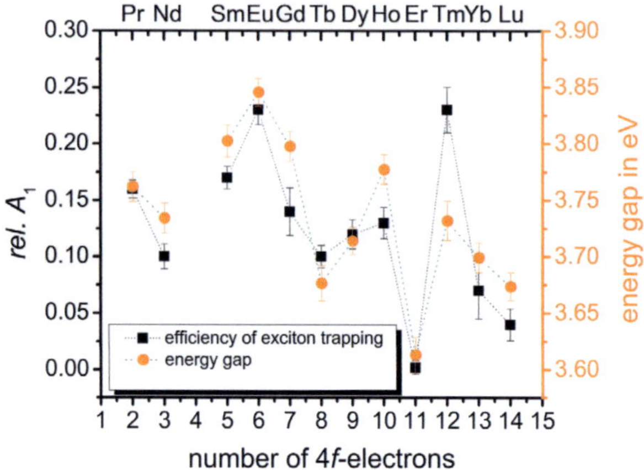

Figure 5.13: Scheme for number of 4f-electrons of trivalent lanthanide ions (exclusive yttrium ions) versus the first relative amplitudes (*rel. A_1*) (square) and the optical energy gap (circle) of iron-lanthanide clusters. Reference **Fe$_6$**-cluster was set as zero point for the relative amplitude axis.

5.8 Photolysis of Iron-lanthanide Clusters

During photoexcitation the pump pulses create a hole nearby the oxygen and cause a transition of a hot electron to the iron ion in the iron oxide domain via ligand to metal charge transfer (LMCT). According to the first relative amplitude, trapping likely occurs at a large fraction of the hot electrons. The trapped electrons then lead to a high electronic density around the iron ion and make the Fe^{3+} ion acting like a pseudo-Fe^{2+} ion (see the picture on the cover of this chapter). A previous study of the absorption spectrum of electrons injected into α-Fe_2O_3 colloidal particles using pulse-radiolysis technique, reported assignment of a broad absorption band extending at least from 500 to 900 nm to electrons trapped at Fe^{3+} sites, reducing them to Fe^{2+}.[316] If "Fe^{2+}" exists after photoexcitation in this cluster system, then this "Fe^{2+}" may be detected using potassium ferricya-

114

nide ($K_3[Fe^{III}(CN)_6]$). In this section, irradiation experiments, including steady-state absorption and transient absorption measurements, were described in details to support this assumption.

5.8.1 Irradiation Experiments

In irradiation experiments, **Fe$_{10}$Er$_{10}$** and **Fe$_{10}$Tm$_{10}$** clusters were chosen because of their opposite properties in color, magnetism and ultrafast dynamics. A mixture of **Fe$_{10}$Er$_{10}$** clusters with potassium ferricyanide was irradiated by a xenon arc lamp (Müller Elektronik-Optik, wavelength spectrum see Appendix C.2). A deep blue color precipitation was formed within 1 second (see Figure 5.14). After a long time of irradiation, a deep blue color precipitation dominates the entire cuvette. The same experimental results were observed in **Fe$_{10}$Tm$_{10}$** clusters (Figure 5.15), as well as in **Fe$_6$**-clusters. In contrast to **Fe$_{10}$Er$_{10}$** and **Fe$_{10}$Tm$_{10}$**, the color change in the **Fe$_6$**-cluster requires a much longer time.[275] Reference samples such as a mixture of the **Dy$_6$**-cluster with potassium ferricyanide, potassium ferricyanide alone, potassium ferricyanide in $Dy(NO_3)_3$, and $Ho(NO_3)_3$ solution, as well as the mixture of **Fe$_{10}$Tm$_{10}$** with potassium ferrocyanide ($K_4[Fe^{II}(CN)_6]$) were prepared under the same conditions for comparison. No color change was observed in these systems. Moreover, direct photolysis of **Fe$_{10}$Ln$_{10}$** clusters gives rise to the destruction of the ring structure and produces a brown suspension. The UV/Vis absorption spectra of **Fe$_{10}$Tm$_{10}$** clusters alone after different irradiation times are shown in Figure 5.16. An additional absorption peaking at 420 nm rises after irradiation indicating the production of iron(III) oxyhydroxide.[317,318] However, the time required to destroy the clusters is much slower than that for the reaction between **Fe$_{10}$Tm$_{10}$** and potassium ferricyanide. A high-pass filter (Schott, WG305) was applied in irradiation measurements to avoid damaging the clusters. Under this condition, the clusters are more stable against irradiation (see Appendix C.3). Consequently, one can safely estimate that the deep blue color arise from the reaction between $Fe_{10}Ln_{10}$ cluster and potassium ferricyanide. This reaction occurs even by excitation with sunlight. In the sunlight, deep blue colloidal substances begin to form in the mixture after a few minutes. However, under darkness, the mixture can be kept clear for months.[275]

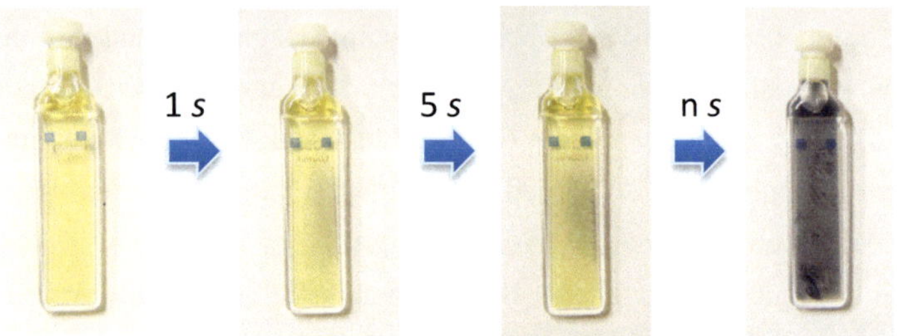

Figure 5.14: Irradiation experiments of the mixture solution containing $Fe_{10}Er_{10}$ clusters and potassium ferricyanide.

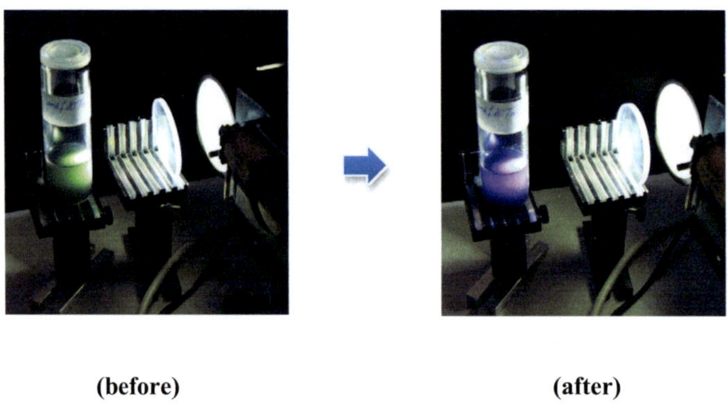

Figure 5.15: Irradiation experiments with filter glass ($\lambda > 270$ nm transmissive) of the mixture solution containing $Fe_{10}Tm_{10}$ clusters and potassium ferricyanide.

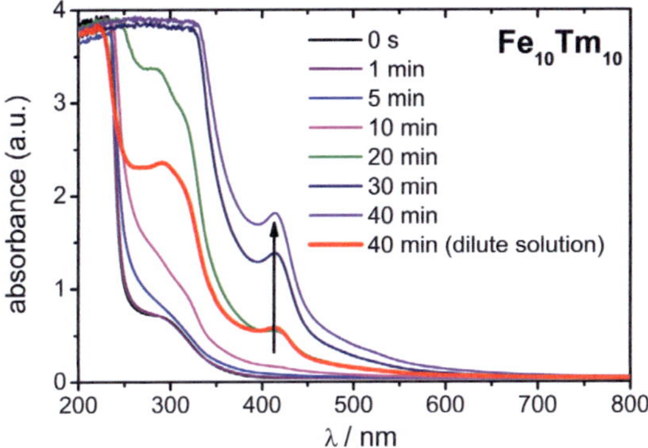

Figure 5.16: UV/Vis absorption spectra of **Fe$_{10}$Tm$_{10}$** clusters after different irradiation times with a xenon arc lamp (measured in a 1 mm cuvette).

5.8.2 UV/Vis Absorption Spectroscopy of the Irradiation Product

The UV/Vis absorption spectrum of the clear filtrate from the deep blue solution is shown in Figure 5.17 in comparison to that of Prussian blue, **Fe$_{10}$Tm$_{10}$** clusters, and potassium ferricyanide. It is well known that Prussian blue as a mixed valence compound with a high spin Fe^{3+} ligated with six N atoms, or a low spin Fe^{2+} ligated with six C atoms of cyanide, can be readily obtained, when Fe^{3+} connects to $[Fe(CN)_6]^{4-}$ or Fe^{2+} to $[Fe(CN)_6]^{3-}$.[319] From the absorption spectrum of Prussian blue, a broad band centered at 690 nm is associated with a metal-metal charge transfer (MMCT) band from Fe^{2+} to Fe^{3+} through a cyanide-ligand bridge.[320,321] The absorption spectrum of the complex from **Fe$_{10}$Tm$_{10}$** with potassium ferricyanide after irradiation, as shown in Figure 5.19a, presents one transition at 310 nm, which can be assigned to ligand metal charge transitions within the iron oxide unit and the transition at 600 nm being assigned to the metal to metal electron transition through the CN-bridge in comparison with the absorption spectrum of Prussian blue at 690 nm. Since the MMCT transition is not possible either from Fe^{3+} to Fe^{3+} or from Fe^{2+} to Fe^{2+}, showing the blue coloration, therefore, the UV/Vis spectrum indicates that the irradiated compound contained an Fe^{3+}-CN-Fe^{2+} unit which bears the Prussian blue analogous spectral properties. The MMCT band in the irradiation product undergoes a blue shift (about 100 nm) to higher energies in comparison with Prussian blue. The absorption spectra of **Fe$_{10}$Tm$_{10}$** clusters alone and potassium ferricyanide did not show any absorption band in the range of 500-800 nm. A new peak at 420 nm may arise from the remaining potassium ferricyanide in solution or from a defected product of the **Fe$_{10}$Tm$_{10}$** cluster after long time irradiation (see Figure 5.16). Photo-induced reaction of the **Fe$_6$**-cluster and potassium ferricyanide can take place under the same conditions. As shown in Figure 5.17b, a broad absorption band exhibits the MMCT transition. (For more details about structural analysis of irradiation product see Ref. [275])

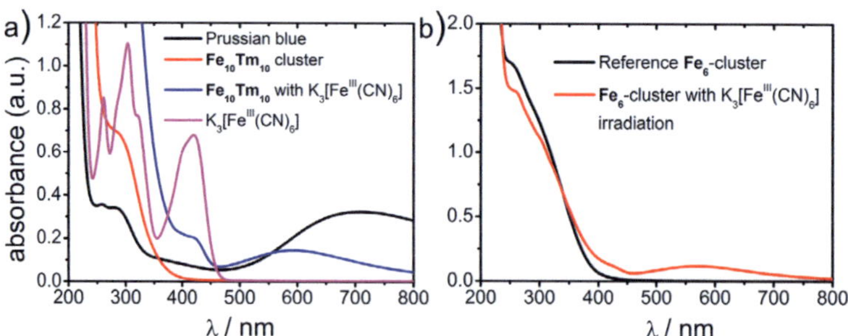

Figure 5.17: a) UV/Vis absorption spectra of Prussian blue (black), **Fe$_{10}$Tm$_{10}$** (red), and **Fe$_{10}$Tm$_{10}$** with K$_3$[Fe^{III}(CN)$_6$] after irradiation with UV-lamp (blue). b) Reference measurement with a mixture of **Fe$_6$**-cluster and potassium ferricyanide. (Measurements were carried out in a 1 mm cuvette).

5.8.3 Pump-probe Spectroscopy of the Irradiation Product

Upon photoexcitation of the compounds into an MMCT band, an electron undergoes an optical transfer process, which can be detected in time-resolved pump-probe measurements.[322-324] The ultrafast dynamics of charge transfer processes on the MMCT band of Prussian blue was studied by Arnett *et al.*[321] In their study, a rapid initial decay from a sub-picosecond to picosecond time scale implied vibronic relaxation and back-electron transfer (BET) from a new product state (MMCT), and a relatively long-lived relaxation suggested to be the lifetime of a charge transfer state (CT-state) on a scale of a few hundred picoseconds. Recent transient absorption studies of Prussian blue, including excitation and probing with both infrared and visible pulses, were reported by Weidinger *et al.* [325]. The result showed that an initial transient bleach which vibrational relaxation in 18 ps was followed by a slower decay on a long (about 370 ps) time scale in IR pump-800 nm probe experiments. The long-lived CT-states have also been suggested to explain the long decay time component in the transient absorption measurements. Moreover, long-lived CT-states in the nanosecond regime were also observed in Prussian blue analogue such as cobalt-containing metal cyanides (Co^{III}-CN-Fe^{II}).[326] For comparison, pump-probe experiments of Prussian blue in water were performed at room temperature after excitation at 310 nm and probed at 550 nm and 650 nm (Figure 5.18). Globe fit of the transients with a tri-exponential function results in three time decay constants: (505 ± 25) fs, (15 ± 5) ps, and (393 ± 16) ps. The third time constant on a few hundred picoseconds is in good agreement with the literature values for the long lifetime of CT-states.[321,325]

The pump-probe measurements of the irradiation product from the mixture of **Fe$_6$**-cluster and potassium ferricyanide show a transient bleaching which is analogue to Prussian blue (Figure 5.19a). The time constants could be determined as (922 ± 63) fs, (9 ± 2) ps, and (266 ± 32) ps, bearing similar decay time scales as those observed in Prussian blue in earlier studies[321,325]. The decay curves measured in the irradiation product of the mixture of **Fe₁₀Tm₁₀** and potassium ferricyanide do not differ from those of the **Fe₆**-cluster with potassium ferricyanide and Prussian blue (Figure 5.19b). A tri-exponential fit up to 50 ps results in three time constants: (565 ± 60) fs, (5 ± 0.6) ps, and $\gg 50$ ps. The measurements with higher delay time than 50 ps were disturbed by newly produced irradiation product in solution. Nevertheless, the long-lived time constant can still be observed.

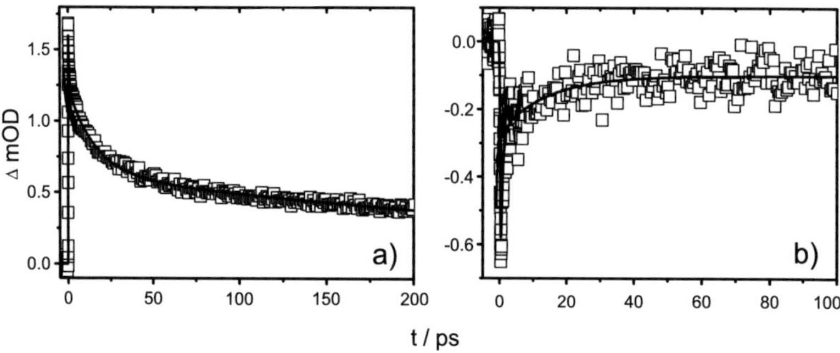

Figure 5.18: Temporal profiles of Prussian blue after excitation at 310 nm and probing at a) 550 nm and b) 650 nm.

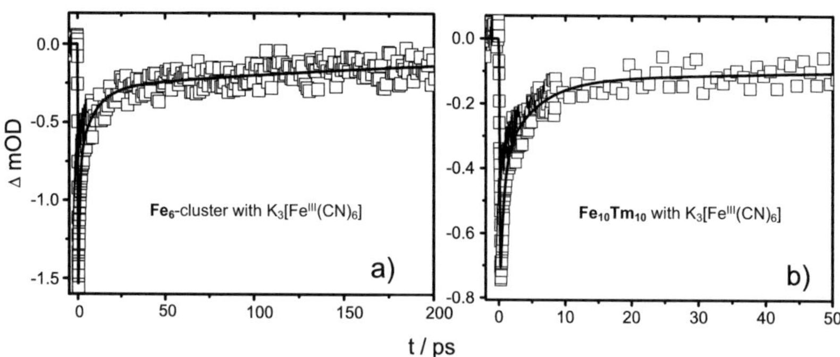

Figure 5.19: Temporal profiles of irradiation products from a) **Fe₆**-cluster and potassium ferricyanide and b) **Fe₁₀Tm₁₀** and potassium ferricyanide after excitation at 310 nm and probing at 550 nm.

5.9 Conclusion and Outlook

In conclusion, the ultrafast dynamics in cyclic coordination iron-lanthanide clusters ($Fe_{10}Ln_{10}$) were investigated applying pump-probe spectroscopic technique in combination with UV/Vis absorption spectroscopy. These clusters showed similar decay profiles of optical responses, which originate from the LMCT transitions within iron-oxygen domain. The fastest decay (200-400 fs) is possibly due to a high density of mid-energy gap trap states induced by doped lanthanide ions, while the second and third decays correspond to electron-hole recombination and the lifetime of trapped exciton, respectively. Due to the high density of trap states, the trapping rates are very rapid and efficient depended on the doped lanthanide ions. In addition, photoreactions between $Fe_{10}Ln_{10}$ clusters and potassium ferricyanide reveal the presence of a pseudo-Fe^{2+} ion produced by photo-excitation. This experimental observation can provide a better understanding of the dynamics. Although the dynamics have been studied thoroughly for the whole trivalent lanthanide ions, the question how lanthanide ions influence the dynamics remains unclear. Further study on this topic will be focused on two aspects: the size-dependent ultrafast dynamics as well as size-dependent ultrafast magnetization dynamics in size-controlled iron-lanthanide clusters.

Appendix

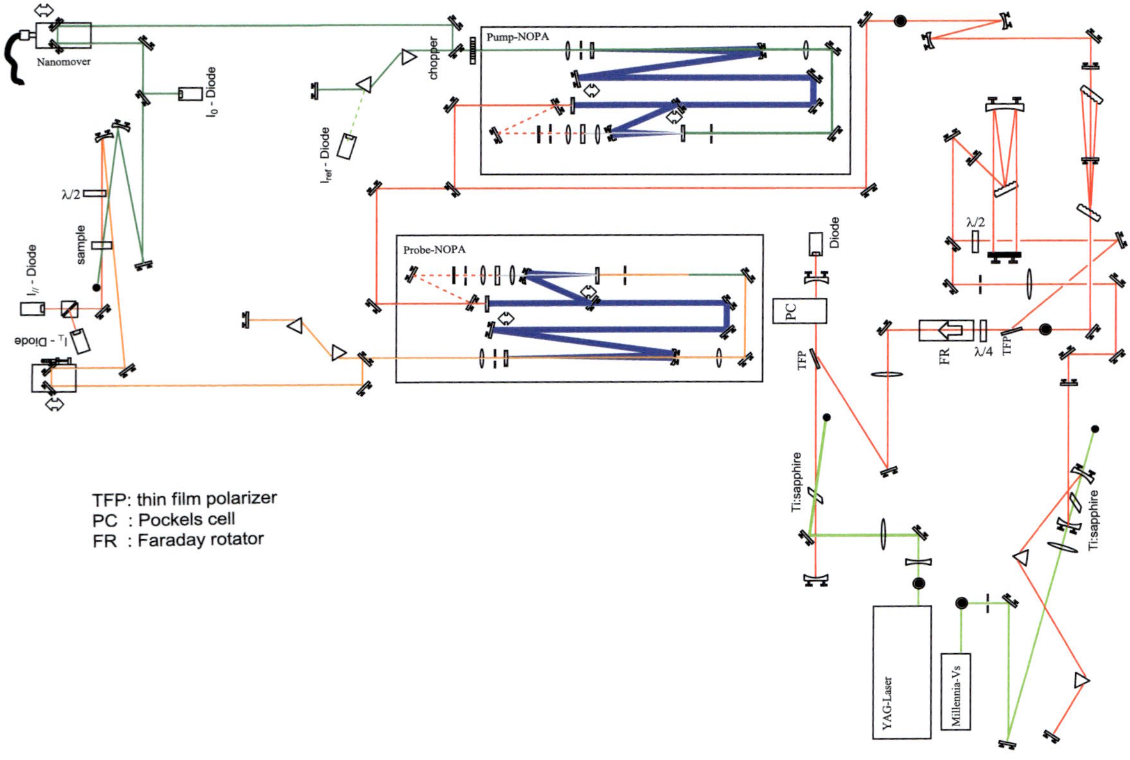

A.1: The home-built femtosecond laser system.

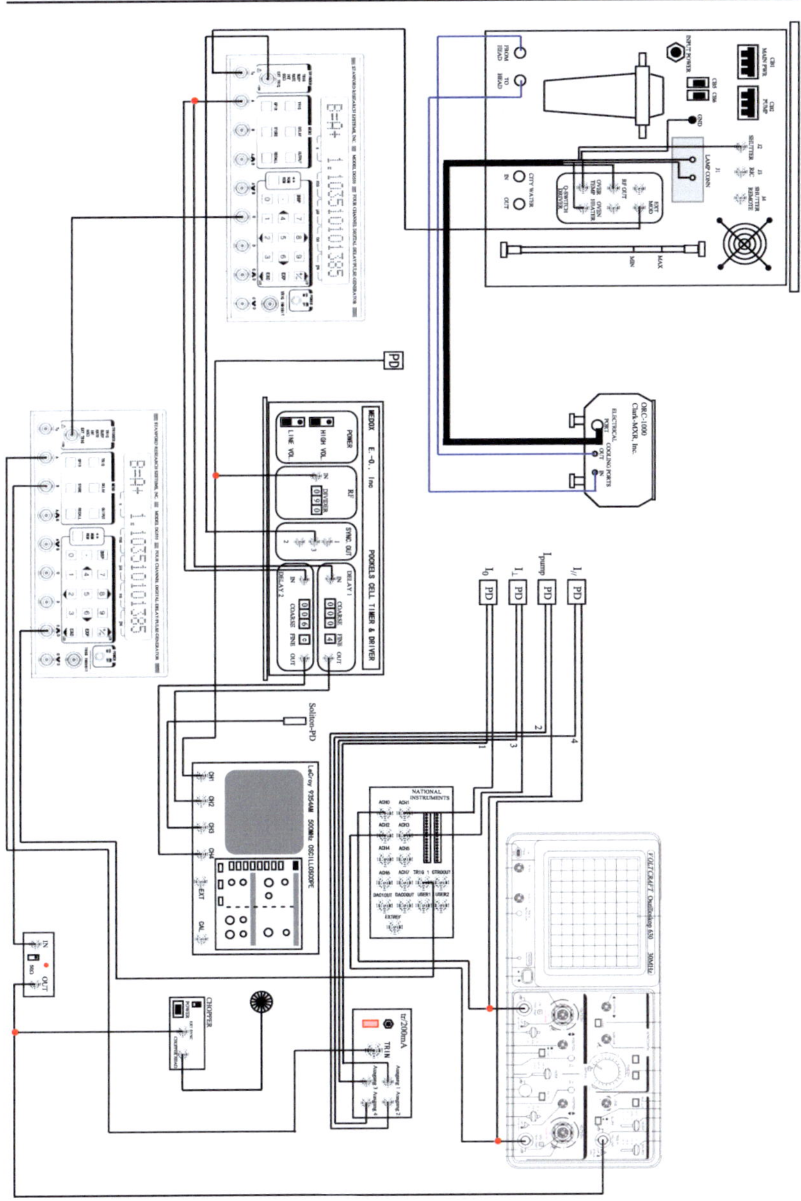

A.2: The associated system for the femtosecond laser system.

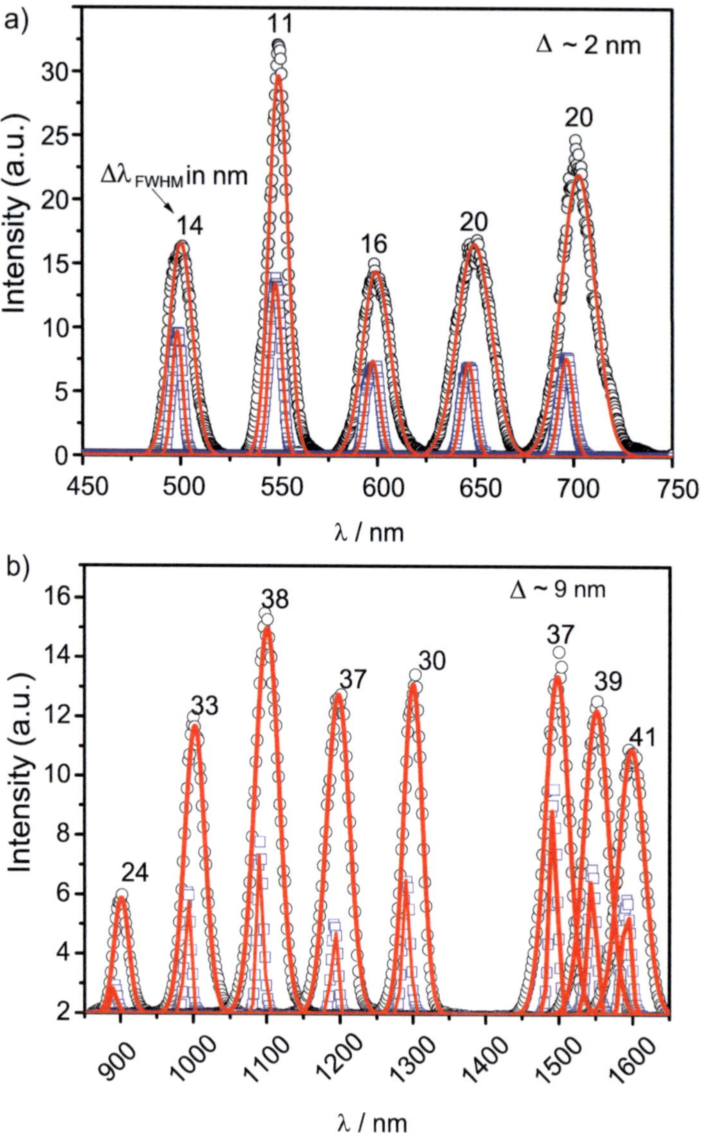

A.3: Calibration of spectral peaks (circle: without filter, square: with filter) in the Vis (a) and NIR (b) regions using the corresponding reference filters (ThorLabs).

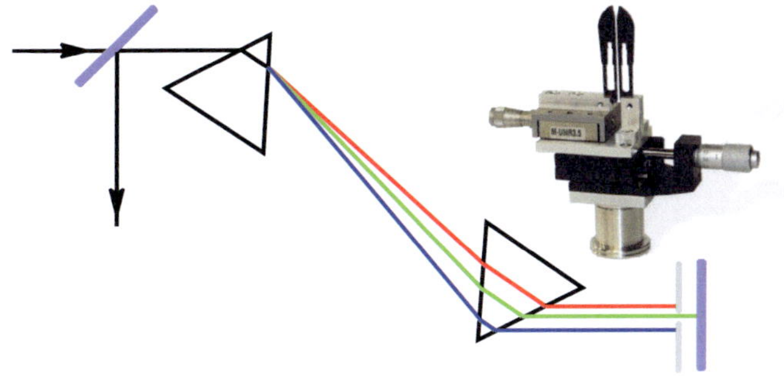

B.1: Spectral narrowing of the probe pulse inside the optical compressor with two variable positioning metal plates.

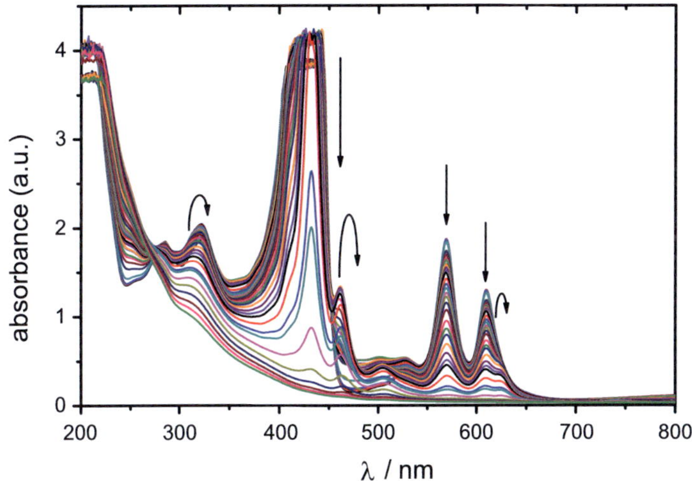

B.2: UV/Vis absorption spectra of CdTPP in THF after different irradiation times using a xenon arc lamp. After 2 hours, the bands completely disappeared. In contrast to CdTPP, MgTPP in THF was more stable under irradiation.

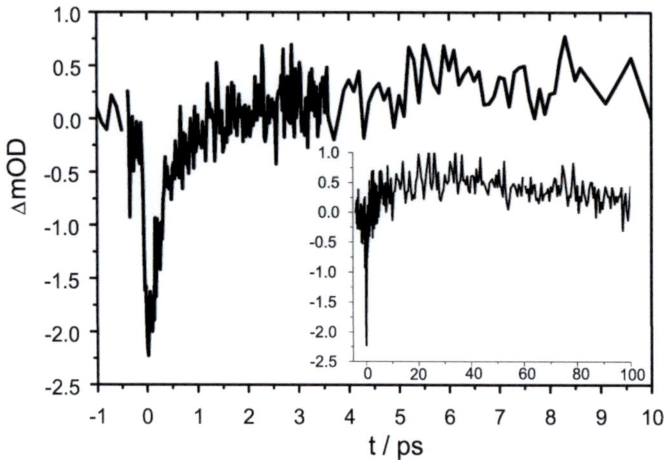

B.3: Transient response of CdTPP in THF after excitation at 610 nm and probing at 615 nm. The measurement was performed in a standing cuvette. The response becomes positive (ΔOD-value) at delay times exceeding 1 ps.

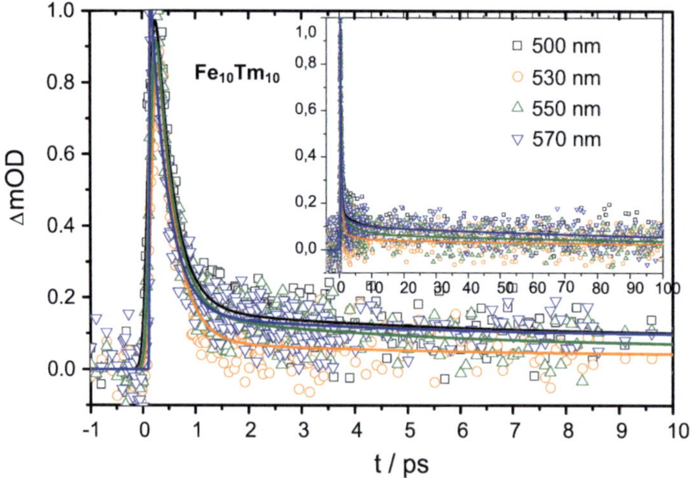

C.1: Probe wavelengths dependence of **Fe$_{10}$Tm$_{10}$** clusters after excitation at 310 nm.

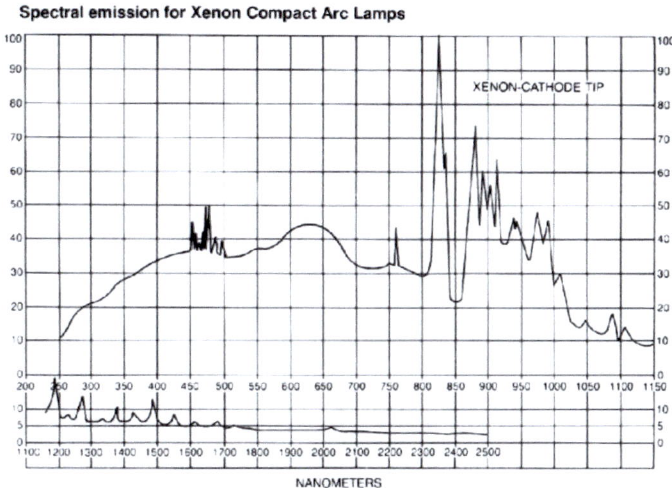

C.2: Wavelength spectrum of a xenon arc lamp (Müller Elektronik-Optik).

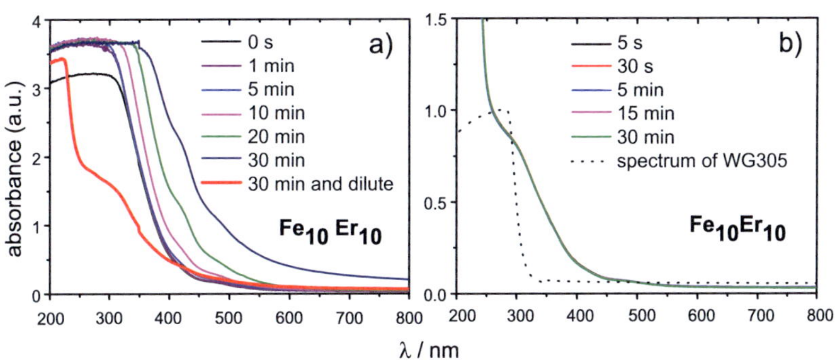

C.3: UV/Vis absorption spectra of **Fe$_{10}$Er$_{10}$** clusters after different irradiation times with a xenon arc lamp (without (a) and with (b) filter glass WG305, $\lambda > 270$ nm transmissive (dot)).

Bibliography

[1]   A. H. Zewail in *Femtochemistry* (Ed.: F. C. De Schryver, S. De Feyter, G. Schweitzer), Wiley-VCH, **2001**.

[2]   C.-H. Chuang, C. Burda, *J. Phys. Chem. Lett.* **2012**, *3*, 1921-1927.

[3]   A. Materny, J. L. Herek, P. Cong, A. H. Zewail, *J. Phys. Chem.* **1994**, *98*, 3352.

[4]   L. R. Milgrom in *The Colors of Life*, Oxford University Press, New York, **1997**.

[5]   X. Liu, E. K. L. Yeow, S. Velate, R. P. Steer, *Phys. Chem. Chem. Phys.* **2006**, *8*, 1298.

[6]   P. Heathcote, P. K. Fyfe, M. R. Jones, *Trends Biochem. Sci.* **2002**, *27*, 79.

[7]   K. Ogawa, T. Zhang, K. Yoshihara, Y. Kobuke, *J. Am. Chem. Soc.* **2002**, *124*, 22-23.

[8]   A. Jabłoński, *Z. Phys.* **1935**, *94*, 38-46.

[9]   P. Klán, J. Wirz in *Photochemistry of Organic Compounds: From Concepts to Practice*, Wiley, **2009**.

[10]   G. Käb, C. Schröder, D. Schwarzer, *Phys. Chem. Chem. Phys.* **2002**, *4*, 271-278.

[11]   L. Luo, C.-H. Chang, Y.-C. Chen, T.-K. Wu, E. W.-G. Diau, *J. Phys. Chem. B* **2007**, *111*, 7656-7664.

[12] A. Gadalla, J.-B. Beaufrand, M. Bowen, S. Boukari, E. Beaurepaire, O. Crégut, M. Gallart, B. Hönerlage, P. Gilliot, *J. Phys. Chem. C* **2010**, *114*, 17854-17863.

[13] M. P. Eng, T. Ljungdahl, J. Andréasson, J. Mårtensson, B. Albinsson, *J. Phys. Chem. A* **2005**, *109*, 1776-1784.

[14] J. D. Le, Y. Pinto, N. C. Seeman, K. Musier-Forsyth, T. A. Taton, R. A. Kiehl, *Nano. Lett.* **2004**, *4*, 2343.

[15] R. D. Felice, D. Porath in *NanoBioTechnology – BioInspired Devices and Materials of the Future* (Ed.: O. Shoseyov, I. Levy), Humana Press, **2008**, pp. 141.

[16] M. Ratner, *Nature*, **1999**, *397*, 480-481.

[17] M. W. Grinstaff, *Angew. Chem. Int. Ed. Engl.* **1999**, *38*, 3629-3635.

[18] G. B. Schuster, *Acc. Chem. Res.* **2000**, *33*, 253.

[19] B. Giese, *Acc. Chem. Res.* **2000**, *33*, 631.

[20] M. E. Nunez, J. K. Barton, *Curr. Opin. Chem. Biol.* **2000**, *4*, 199.

[21] M. Bixon, J. Jortner, *Chem. Phys.* **2002**, *281*, 393.

[22] F. D. Lewis, R. L. Letsinger, M. R. Wasielewski, *Acc. Chem. Res.* **2001**, *34*, 159.

[23] S. M. J. Aubin, M. W. Wemple, D. M. Adams, H.-L. Tsai, G. Christou, D. N. Hendrickson, *J. Am. Chem. Soc.* **1996**, *118*, 7746.

[24] M. Mannini, F. Pineider, P. Sainctavit, C. Danieli, E. Otero, C. Sciancalepore, A. M. Talarico, M. A. Arrio, A. Cornia, D. Gatteschi, R. Sessoli, *Nat. Matter.* **2009**, *8*, 194.

[25] A. Baniodeh, I. J. Hewitt, V. Mereacre, Y.-H. Lan, G. Novitchi, C. E. Anson, A. K. Powell, *Dalton Trans.* **2011**, *40*, 4080.

[26] V. Mereacre, A. Baniodeh, C. E. Anson, A. K. Powell, *J. Am. Chem. Soc.* **2011**, *133*, 15335-15337.

[27] A. M. Ako, V. Mereacre, R. Clérac, W. Wernsdorfer, I. J. Hewitt, C. E. Anson, A. K. Powell, *Chem. Commun.* **2009**, 544.

[28] S. Osa, T. Kido, N. Matsumoto, N. Re, A. Pochaba, J. Mrozinski, *J. Am. Chem. Soc.* **2003**, *126*, 420-421.

[29] C. M. Zaleski, E. C. Depperman, J. W. Kampf, M. L. Kirk, V. L. Pecoraro, *Angew. Chem. Int. Ed.* **2004**, *43*, 3912-3914.

[30] C. Benelli, D. Gatteschi, *Chem. Rev.* **2002**, *102*, 2369-2387.

[31] M. Murugesu, A. Mishra, W. Wernsdorfer, K. A. Abboud, G. Christou, *Polyhedron*, **2006**, *25*, 613-625.

[32] A. Mishra, W. Wernsdorfer, S. Parsons, G. Christou, E. K. Brechin, *ChemComm.* **2005**, 2086-2088.

[33] A. M. Ayuk, V. Mereacre, Y.-H. Lan, W. Wernsdorfer, R. Clérac, C. E. Anson, A. K. Powell, *Inorg. Chem.* **2010**, *49*, 1-3.

[34] W. Schmitt, J. P. Hill, M. P. Juanico, A. Caneschi, F. Constntino, C. E. Anson, A. K. Powell, *Angew. Chem. Int. Ed.* **2005**, *117*, 4173.

[35] M. Andruh, I. Ramade, E. Codjovi, O. Guillou, O. Kahn, J. C. Trombe, *J. Am. Chem. Soc.* **1993**, *115*, 1822.

[36] I. Ramade, O. Kahn, Y. Jeannin, F. Robert, *Inorg. Chem.* **1997**, *36*, 930.

[37] E. Cremades, S. Gómez-Coca, D. Aravena, S. Alvarez, E. Ruiz, *J. Am. Chem. Soc.* **2012**, *134*, 10532-10542.

[38] D. Fleisch in *A Student's Guide to Maxwell's Equations*, Cambridge University Press, **2008**.

[39] D. L. Mills in *Nonlinear Optics – Basic Concepts*, Springer-Verlag, Berlin Heidelberg, **1991**, pp. 4.

[40] P. A. Franken, A. E. Hill, C. W. Peter, G. Weinreich, *Phys. Rev. Lett.* **1961**, 7, 118.

[41] M. Bass, P. A. Franken, J. F. Ward, G. Weinreich, *Phys. Rev. Lett.* **1962**, 9, 446.

[42] M. Bass, P. A. Franken, J. F. Ward, *Phys. Rev.* **1965**, *138*, A 534.

[43] R. W. Boyd in *Nonlinear Optics* 3rd, Academic Press, San Diego, **2007**.

[44] L. A. Golovan, V. Yu. Timoshenko, A. B. Fedotov, L. P. Kuznetsova, D. A. Sidorov-Biryukov, P. K. Kashkarov, A. M. Zheltikov, D. Kovalev, N. Künzner, E. Gross, J. Diener, G. Polisski, F. Koch, *Appl. Phys. B* **2001**, *73*, 31-34.

[45] R. W. Terhune, M. Nisenoff, C. M. Savage, *Phys. Rev. Lett.* **1962**, *8*, 21.

[46] J. A. Giordmaine, *Phys. Rev. Lett.* **1962**, *8*, 19.

[47] M. Abramowitz, M. Davidson in *Optical Birefringence*, Olympus American Research, USA, **1991**.

[48] A. Zheltikov, A. L'Huillier, F. Krausz in *Handbook of Lasers and Optics* (Ed.: F. Träger) Springer, New York, **2007**, pp. 161.

[49] J.-Q. Yao, Y. Wang in *Nonlinear Optics and Solid-State Lasers*, Springer, Berlin, **2012**, pp. 90.

[50] A.-N. Unterreiner in *Ultraschnelle Relaxationsdynamik Solvatisierter Elektronen in Flüssigem Ammoniak und Wasser*, Universität Karlsruhe (TH), Verlag Mainz, **1988**, ISBN: 3-89653-368-1.

[51] C. Rullière in *Femtosecond Laser Pulses: Principles and Experiments* 2nd, **2005**, pp. 46.

[52] G. New in *Introduction to Nonlinear Optics*, Cambridge, **2011**, pp. 133.

[53] A. E. Siegman in *Lasers*, University Science Books, Mill Valley, **1986**.

[54] T. Brabec, C. Spielmann, P. F. Curley, F. Krausz, *Opt. Lett.* **1992**, *17*, 1292-1294.

[55] H. Abramczyk in *Introduction to Laser Spectroscopy*, Elsevier, **2005**, pp. 152-156.

[56] J. Kleinbauer, R. Knappe, R. Wallenstein in *Femtosecond Technology for Technical and Medical Applications* (Ed.: F. Dausinger, F. Lichtner, H. Lubatschowski), Springer Berlin Heidelberg, **2004**, pp. 25.

[57] S. R. Greenfield, M. R. Wasielewski, *Appl. Opt.* **1995**, *34*, 2688.

[58] A. Shirakawa, I. Sakane, M. Takasaka, T.Kobayashi, *Appl. Phys.Lett.* **1999**, *74*, 2268-2270.

[59] T. Wilhelm, J. Piel, E. Riedle, *Opt. Lett.* **1997**, *22*, 1494.

[60] E. Rielde, M. Beutter, S. Lochbrunner, J. Piel, S. Schenkl, S. Spörlein, W. Zinth, *Appl. Phys. B* **2000**, *71*, 457.

[61] T. Wolf in *Ultrafast Photophysical and Photochemical of Radical Precursors in Solution*, Dissertation, Karlsruher Institute of Technology (KIT), **2012**.

[62] M. Boulton, *J. Photochem.&Photobio. B: Biology*, **2001**, *64*, 144.

[63] H. Shinmori, T.Kajiwara, A.Osuka, *Tetrahedron* **2001**, *42*, 3617.

[64] A. Kay, R. Humphry-Baker, M. Grätzel, *J. Phys. Chem.* **1994**, *98*, 952.

[65] D. Wu, Z. Shen, Z. Xue, M.You, *Chin.J. Inorg. Chem.* **2007**, *23*, 1.

[66] E. D. Sternberg, D. Dolphin, C. Bruckner, *Tetrahedron* **1998**, *54*, 4151.

[67] X. Chen, C. M. Drain, *Drug Design Rev.* **2004**, *1*, 215.

[68] M. P. Donzello, C. Ercolani, P. A. M. P. Stuzhin, *Coord. Chem. Rev.* **2006**, *250*, 1530.

[69] V. Gottumukkala, R. J. Luguya, F. R. Fronczeck, M. G. H. Vincente, *Bioorg. Med. Chem.* **2005**, *13*, 1633.

[70] C. A. Mirkin, M. A. Ratner, *Annu. Rev. Phys. Chem.* **1992**, *43*, 719.

[71] M. J. Gunter, M. R. Johnston, *Chem. Soc., Chem. Commun.* **1992**, *17*, 1163.

[72] K. Sienicki in *Molecular Electronics and Molecular Electronic Devices*, CRC Press, **1994**, Vol. III.

[73] J. Karolczak, D. Kowalska, A. Lukaszewicz, A. Maciejewski, R. P. Steer, *J. Phys. Chem. A* **2004**, *108*, 4570.

[74] H. Z. Yu, J. S. Baskin, B. Steiger, C. Z. Wan, F. C. Anson, A. H. Zewail, *Chem. Phys. Lett.* **1998**, *293*, 1.

[75] R. E. Haddad, S. Gazeau, J. Pécaut, J.-C.Marchon, C. J. Medforth, J. A. Shelnutt, *J. Am. Chem. Soc.* **2003**, *125*, 1253.

[76] A. Marcelli, P. Foggi, L. Moroni, C. Gellini, P. R. Salvi, *J. Phys. Chem. A* **2008**, *112*, 1864.

[77] A. Hazell, *ActaCryst.* **1986**, *C42*, 296-299.

[78] A. Hazell, *ActaCryst.* **1984**, *C40*, 751-753.

[79] J. L. Hoard in *Porphyrins and Metalloporphyrins* (Ed.: K. M. Smith) Elsevier, Amsterdam, **1975**, 317-380.

[80] T. A. Evans, J. J. Katz, *Biochimica et Biophysica Acta*, **1975**, *396*, 414-426.

[81] H. Z. Yu, J. S. Baskin, A. H. Zewail, *J. Phys. Chem. A* **2002**, *106*, 9845.

[82] N. Mataga, Y. Shibata, H. Chosrowjan, N. Yoshida, A. Osuka, *J. Phys. Chem. B* **2000**, *104*, 4001.

[83] U. Tripathy, D. Kowalska, X. Liu, S. Velate, R. P. Steer, *J. Phys. Chem. A* **2008**, *112*, 5824.

[84] X. Liu, U. Tripathy, S. V. Bhosale, S. J. Langford, R. P. Steer, *J. Phys. Chem. A* **2008**, *112*, 8986.

[85] S. Sorgues, L. Poisson, K. Raffael, L. Krim, B. Soep, N. Shafizadeh, *J. Chem. Phys.* **2006**, *124*, 114302.

[86] A. Lukaszewicz, J. Karolczak, D. Kowalska, A. Maciejewski, M. Ziolek, R. P. Steer, *Chem. Phys.* **2007**, *331*, 359.

[87] S. Velate, X. Liu, R. P. Steer, *Chem. Phys. Lett.* **2006**, *427*, 295.

[88] J. S. Baskin, H. Z. Yu, A. H. Zewail, *J. Phys. Chem. A* **2002**, *106*, 9837.

[89] X. Liu, E. K. L. Yeow, S. Velate, R. P. Steer, *Phys. Chem. Chem. Phys.* **2006**, *8*, 1298.

[90] G. Niranjian, V. Marat, J. Lasse, K. Karol, *J. Phys. Chem. A* **2009**, *113*, 6041.

[91] Z.-L. Cai, M. J. Crossley, J. R. Reimers, R. Kobayashi, R. D. Amos, *J. Phys. Chem. B* **2006**, *110*, 15624.

[92]   R. Kobayashi, R. D. Amos, *Chem. Phys. Lett.* **2006**, *420*, 106.

[93]   O. Schalk, H. Brands, T. S. Balaban, A.-N. Unterreiner, *J. Phys. Chem. A* **2008**, *112*, 1719.

[94]   Y. Liang, M. Bradler, M. Klinger, O. Schalk, M. C. Balaban, T. S. Balaban, E. Riedle, A.-N. Unterreiner, *ChemPlusChem.* **2013**, to be submitted.

[95]   M. Bradler, P. Baum, E. Riedle, *Appl. Phys. B* **2009**, *97,* 561.

[96]   U. Megerle, I. Pugliesi, C. Schriever, C. F. Sailer, E. Riedle, *Appl. Phys B* **2009**, *96*, 215.

[97]   M. Gouterman in *The Porphyrins* (Ed.: D. Dolphin) Academic Press, New York, **1978**, Vol. 3, p.1.

[98]   K. Kalyanasundaram in *Photochemistry of Polypyridine and Porphyrin Complexes,* Academic Press, London, **1992**.

[99]   V. V. Apanasovich, E. G. Novikov, N. N. Yatskov, R. B. M. Koehorst, T. J. Schaafsma, A. Van Hoek, *J. Appl. Spectrosc.* **1999**, *66(4)*, 613.

[100]   S. F. Shkirman, K. N. Solov'ev, T. F. Kachura, S. A. Arabei, E. D. Skakovskii, *J. Appl. Spectrosc.* **1999**, *66(1)*, 68.

[101]   I. T. Oliver, W. A. Rawlinson, *Biochem. J.* **1955**, *61(4)*, 641.

[102]   J.-L. Soret, *Comptes rendus de l'Académie des sciences*, **1883**, *97*, 1269-1273.

[103]   L. Edwards, D. H. Dolphin, M. Gouterman, *J. Mol. Spectrosc.* **1971**, *38*, 16-32.

[104]   M. Gouterman, F. P. Scharz, P. D. Smith, *J. Chem. Phys.* **1973**, *59*, 676-690.

[105] L. Edwards, D. H. Dolphin, M. Gouterman, *J. Mol. Spectrosc.* **1970**, *35*, 90.

[106] M. Gouterman, *J. Chem. Phys.* **1959**, *30*, 1139.

[107] M. Gouterman, *J. Mol. Spectrosc.* **1961**, *6*, 138.

[108] C. Weiss, H. Kobayashi, M. Gouterman, *J. Mol. Spectrosc.* **1965**, *16*, 415.

[109] E. J. Baerends, G. Ricciardi, A. Rosa, S. J. A. van Gisbergen, *Coord. Chem. Rev.* **2002**, *230*, 5.

[110] H. Solheim, K. Ruud, S. Coriani, P. Norman, *J. Phys. Chem .A* **2008**, *112*, 9615.

[111] G. A. Peralta, M. Seth, T. Ziegler, *Inorg. Chem.* **2007**, *46*, 9111.

[112] L. R. Milgrom *in The Colours of Life: An Introduction on the Chemistry of Porphyrins and Related Compounds*, Oxford, **1997**.

[113] K. S. Suslick, R. A. Waston, *new J. Chem.* **1992**, *16*, 633-642.

[114] R. M. Wang, B. M. Hoffman, *J. Am. Chem. Soc.* **1984**, *106*, 4235.

[115] N. S. Venkatamanan, A. Suvitha, H. Nejo, H. Mizuseki, Y. Kawazoe, *Int. J. Quant. Chem.* **2011**, *111*, 2340-2351.

[116] R. Fukuda, M. Ehara, H. Nakatsuji, *J. Chem. Phys.* **2010**, *133*, 144316.

[117] A. Dedieu, M.-M. Rohmer, A. Veillard, *Adv. Quantum Chem.* **1982**, *16*, 43-95.

[118] M. Rubio, B. O. Roos, L. Serrano-Andrés, M. Merchán, *J. Chem. Phys.* **1999**, *110*, 7202-7209.

[119] P. Suppan, *J. Photochem. Photobiol. A: Chem.* **1990**, *50*, 293-330.

[120] C. Reichardt, T. Welton in *Solvents and Solvent Effects in Organic Chemistry* 3rd , **2003**, pp. 360.

[121] O. Schalk, Y. Liang, A.-N. Unterreiner, *Z. Phys. Chem.* **2013**, *227*, 35-48.

[122] T. S. Balaban, P. Braun, C. Hättig, A. Hellweg, J. Kern, W. Saenger, A. Zouni, *Biochimica Biophysica Acta, Bioenergetics*, **2009**, *1787*, 1254-1265.

[123] T. Oba, H. Tamiaki, *Photosynth. Res.* **2002**, *74*, 1.

[124] T. S. Balaban, P. Fromme, A. R. Holzwarth, N. Krauß, V. I. Prokhorenko, *BiochimiaBiophysicaActa, Bioenergetics*, **2002**, *1556*, 197.

[125] L. L. Shipman, T. M. Cotton, J. R. Norris, J. J. Katz, *Proc. Natl. Acad. Sci. USA* **1976**, *73*, 1791.

[126] A. Ben Fredj, Z. Ben Lakhdar, M. F. Ruiz-López, *Chem. Phys. Lett.* **2009**, *472*, 243-247.

[127] W.-S Wun, J.-H. Chen, S.-S. Wang, J.-Y. Tung, F.-L. Liao, S.-L. Wang, L.-P. Hwang, S. Elango, *Inorg. Chem. Commun.* **2004**, *7*, 1233-1237.

[128] B. Minaev, Y.-H.Wang, C.-K. Wang, Y. Luo, H. Ågren, *Spectrochim. Acta. Part A* **2006**, *65*, 308-323.

[129] H. Chosrowjan, S. Tanigichi, T. Okada, S. Takagi, T. Arai, K. Tokumaru, *Chem. Phys. Lett.* **1995**, *242*, 644.

[130] G. G. Gurzadyan, T.-H Tran-Thi, T. Gustavsson, *J. Chem. Phys.* **1998**, *108*, 385.

[131] D. Kowalska, R. P. J. Steer, *Photochem. Photobiol. A* **2008**, *195*, 223.

[132] A. Harriman, *J. Chem. Soc., Faraday Trans. 2* **1981**, *77*, 1281.

[133] P. G. Seybold, M. Gouterman, *J. Mol. Spectrosc.* **1969**, *31*, 1-13.

[134] M. Kasha, H. R. Rawls, M. Ashraf El-Bayoumi, *Pure Appl. Chem.* **1965**, *11*, 371-392.

[135] R. Lauceri, M. De Napoli, A. Mammana, S. Nardis, A. Romeo, R. Purrello, *Synth. Met.* **2004**, *147*, 49-55.

[136] P. S. Santiago, S. C. M. Gandini, L. M. Moreira, M. Tabak, *J. Porphyrins Phthalocyanines,* **2008**, *12*, 942-952.

[137] K. Hosomizu, M. Oodoi, T. Umeyama, Y. Matano, K. Yoshida, S. Isoda, M. Isosoppi, N. V. Tkachenko, H. Lemmetyinen, H. Imahori, *J. Phys. Chem. B* **2008**, *112*, 16517-16524.

[138] A. H. Herz, *Adv. Colloid Interfac.* **1977**, *8*, 237-298.

[139] W. I. White in *The Porphyrins* (Ed.: D. Dolphin), Academic Press, New York, **1978**, pp. 303.

[140] E. G. Mc Rae, M. Kasha, *J. Chem. Phys.* **1958**, *28*, 721.

[141] D. Kim, A. Osuka, *J. Phys. Chem. A* **2003**, *107*, 8791-8816.

[142] H. S. Cho, N. W. Song, Y. H. Kim, S. C. Jeoung, S. Hahn, D. Kim, S. K. Kim, N. Yoshida, A. Osuka, *J. Phys. Chem. A* **2000**, *104*, 3287-3298.

[143] J. Yang, D. Kim, *Phil. Trans. R. Soc. A* **2012**, *370*, 3802-3818.

[144] J. Rodriguez, C. Kirmaier, D. Holten, *J. Am. Chem. Soc.* **1989**, *111*, 6501.

[145] G. E. O'Keefe, G. J. Denton, E. J. Harvey, R. T. Phillips, R. H. Friend, H. L. Anderson, *J. Chem. Phys.* **1996**, *104* 805.

[146] G. G. Gurzadyan, T.-H.Tran-Thi, T. Gustavsson, *J. Chem. Phys.* **1998**, *108*, 385.

[147] E. G. Azenha, A. C. Serra, M. Pineiro, M. M. Pereira, J. Seixas de Melo, L. G. Arnaut, S. J. Formosinho, A. M. d'A. Rocha Gonsalves, *Chem. Phys.* **2002**, *280*, 177.

[148] D. Wróbel, J. Łukasiewicz, H. Manikowski, *Dyes and Pigments* **2003**, *58*, 7.

[149] C. Galli, K. Wynne, S. M. LeCours, M. J. Therien, R. M. Hochstrasser, *Chem. Phys. Lett.* **1993**, *206*, 493.

[150] P. S. Zhao, F. F. Jian, L. Zhang, *Bull. Korean Chem. Soc.* **2006**, *27*, 1053-1055.

[151] M. Z. Zgierski, T. Fujiwara, E. Lim, *Acc. Chem. Res.* **2010**, *43*, 506.

[152] Y. Kurabayashi, K. Kikuchi, H. Kokubum, Y. Kaizu, H. Kobayashi, *J. Phys. Chem.* **1984**, *88*, 1308.

[153] O. Ohno, Y. Kaizu, H. Kobayashi, *J. Chem. Phys.* **1985**, *82*, 1779.

[154] Y. Kaizu, M. Asano, H. Kobayashi, *J. Phys. Chem.* **1986**, *90*, 3906.

[155] N. J. Turro, V. Ramamurthy, J. Scaiano in *Modern Molecular Photochemistry of Organic Molecules*, University Science Books, New York, **1991**.

[156] B. Valeur, in *Molecular Fluorescence: Principles and Applications*, Wiley-VCH, Weinheim, **2001**.

[157] S. S. Vogel, C. Thaler, P. S. Blank, S. V. Koushik in *FLIM Microscopy in Biology and Medicine* (Ed.: A. Periasamy, R. M. Clegg), Chapman & Hall/CRC, **2009**, pp. 245-288.

[158] A. Kawski, *Crit. Rev. Anal. Chem.* **1993**, *23*, 459.

[159] H. J. Bakker, J. Woutersen, H.-K. Nienhuys, *Chem. Phys.* **2000**, *258*, 233.

[160] A. T. Yeh, C. V. Shank, J. K. McCusker, *Science*, **2000**, *289*, 935-938.

[161] K. Wynne, R. M. Hochstrasser, *Chem. Phys.* **1993**, *171*, 179.

[162] H. Hippler, M. Olzmann, O. Schalk, A.-N. Unterreiner, *Z. Phys. Chem.* **2005**, *219*, 389-398.

[163] R. S. Knox, R. C. Gilmore, *J. Lumin.* **1995**, *63*, 163.

[164] S. Raghavan, R. S. Birch, J. H. Eberly, *Chem. Phys. Lett.* **2000**, *326*, 207.

[165] D. A. Farrow, E. R. Smith, W. Qian, D. M. Jonas, *J. Chem. Phys.* **2009**, *128*, 144510.

[166] E. R. Smith and D. M. Jonas, *J. Phys. Chem. A* **2011**, *115*, 4101.

[167] O. Schalk and A.-N. Unterreiner, *Z. Phys. Chem.* **2011**, *225*, 927.

[168] K. Wynne, R. M. Hochstrasser, *J. Raman Spectrosc.* **1995**, *26*, 561-569.

[169] Y. Liang, O. Schalk, A.-N. Unterreiner, *EPJ Web of Conferences*, **2013**, *41*, 05014.

[170] O. Schalk, A.-N. Unterreiner, *Phys. Chem. Chem. Phys.* **2010**, *12*, 655-666.

[171] N. F. Scherer, L. R. Khundkar, T. S. Rose, A. H. Zewail, *J. Phys. Chem.* **1987**, *91*, 6478.

[172] F. Perrin, *J. Phys. Radium* **1934**, *5*, 497.

[173] S. Hess, H. Bürsing, P. Vöhringer, *J. Chem. Phys.* **1999**, *111*, 5461.

[174] J. D. Watson and F. H. C. Crick, *Nature*, **1953**, *171*, 737.

[175] R. Franklin, R. G. Gosling, *Nature*, **1953**, *171*, 740.

[176] R. E. Holmlin, P. J. Dandliker, J. K. Barton, *Angew. Chem. Int. Ed.* **1997**, *36*, 2714.

[177] H. Rozenberg, D. Rabinovich, F. Frolow, R. S. Hegde, Z. Shakked, *Proc. Natl. Acad. Sci. USA* **1998**, *95*, 15194-15199.

[178] W. A. Bernhard, D. M. Close in *Charged Particle and Proton Interactions with Matter* (Eds.: A. Mozumder, Y. Hatano) Marcel Dekker New York, **2004**.

[179] G. Kraft, M. Kramer, *Adv. Radiation Biol.* **1993**, *17*, 1-52.

[180] R. A. Jr. Casero, P. Celano, S. J. Ervin, N. B. Applegren, L. Wiest, A. E. Pegg, *J. Biol. Chem.* **1991**, *266*, 810-814.

[181] T. A. Jacob, P. Jaruga, R. G. Nath, M. Dizdaroglu, P. J. Brooks, *Nucleic Acids Res.* **2005**, *33*, 3513-3520.

[182] M. Kawakita, K. Hiramatsu, *J. Biochem.* **2006**, *139*, 315-322.

[183] L. Alhonen, J. J. Parkkinen, T. Keinanen, *Proc. Natl. Acad. Sci. USA* **2000**, *97*, 8290-8295.

[184] E. Meggers, M. E. Michel-Beyerle, B. Giese, *J. Am. Chem. Soc.* **1998**, *120*, 12950-12955.

[185] P. T. Henderson, D. Jones, G. Hampikian, Y. Z. Kan, G. B. Schuster, *Proc. Natl. Acad. Sci. USA* **1999**, *96*, 8353-8358.

[186] D. D. Eley, D. I. Spivey, *Trans. Faraday Soc.* **1962**, *58*, 411-415.

[187] C. R. Calladine, H. R. Drew, B. F. Luisi in *Understanding DNA – The Molecule & How It Works* 3rd, Elsevier, **2004**.

[188] S. O. Kelly, J. K. Barton, *Science* **1999**, *283*, 375-381.

[189] S. O. Kelly, N. M. Jackson, M. G. Hill, J. K. Barton, *Angew.Chem. Int. Ed. Engl.* **1998**, *38*, 941.

[190] C. Behrens, M. K. Cichon, F. Grolle, U. Hennecke, T. Carell in*Topics in Current Chemistry: Long-Range Electron Transfer in DNA I*, (Eds.: G. B. Schuster) Springer New York, **2004**, pp. 187-204.

[191] B. Giese, J. Amaudrut, A.-K. Köhler, M. Spormann, S. Wessely, *Nature*, **2001**, *412*, 318-320.

[192] B. Giese, *Top.Curr.Chem.* **2004**, *236*, 27-44.

[193] F. D. Lewis, H. Zhu, P. Daublain, B. Cohen, M. R. Wasielewski, *Angew. Chem. Int. Ed.* **2006**, *45*, 7982-7985.

[194] T. Shimazaki, Y. Asai, K. Yamashita, *J. Phys. Chem.* **2005**, *109*, 1295-1303.

[195] F. D. Lewis, X. Liu, J. Liu, S. E. Miller, R. T. Hayes, M. R. Wasielewski, *Nature*, **2000**, *406*, 51-53.

[196] S. M. M. Conron, A. K. Thazhatthveetil, M. R. Wasielewski, A. L. Burin, F. D. Lewis, *J. Am. Chem. Soc.* **2010**, *132*, 14388-14390.

[197] M. J. Park, M. Fujitsuka, K. Kawai, T. Majima, *J. Am. Chem. Soc.* **2011**, *133*, 15320-15323.

[198] T. Takada, K. Kawai, M. Fujitsuka, T. Majima, *Chem. Eur. J.* **2005**, *11*, 3835-3842.

[199] F. D. Lewis, R. L. Letsinger, M. R. Wasielewski, *Acc. Chem. Res.* **2001**, *34*, 159.

[200] B. Giese, *Ann. Rev. Biochem.* **2002**, *71*, 51.

[201] S. Delaney, J. K. Barton, *J. Org. Chem.* **2003**, *68*, 6475.

[202] G. B. Schuster, *Acc. Chem. Res.* **2000**, *33*, 253.

[203] Z. Cai, M. D. Sevilla, *J. Phys. Chem. B* **2000**, *104*, 6942.

[204] E. Hatcher, A. Balaeff, S. Keinan, R. Venkatramani, D. N. Beratan, *J. Am. Chem. Soc.* **2008**, *130*, 17752-17761.

[205] F. C. Grozema, S. Tonzani, Y. A. Berlin, G. C. Schatz, L. D. Siebbeles, M. A. Ratner, *J. Am. Chem. Soc.* **2008**, *130*, 5157-5166.

[206] T. Kubar, U. Kleinekathöfer, M. Elstner, *J. Phys. Chem. B* **2009**, *113*, 13107-13117.

[207] F. C. Grozema, Y. A. Berlin, L. D. A. Siebbeles, *J. Am. Chem. Soc.* **2000**, *122*, 10903-10909.

[208] M. Bixon, J. Jortner, *J. Am. Chem. Soc.* **2001**, *123*, 12556-12567.

[209] V. D. Lakhno, V. B. Sultanov, *Biophysics* **2007**, *52*, 24-29.

[210] T. Ehrenschwender, Y. Liang, A.-N. Unterreiner, H.-A. Wagenknecht, T. J. A. Wolf, *ChemPhysChem.* **2013**, *14*, 1197-1204.

[211] J. Genereux, J. K. Barton, *Chem. Rev.* **2010**, *110*, 1642-1662.

[212] F. D. Lewis, M. R. Wasielewski, *Top. Curr. Chem.* **2004**, *236*, 45-65.

[213] K. Kawai, T. Majima, *Top. Curr. Chem.* **2004**, *236*, 117-137.

[214] G. B. Schuster, U. Landman, *Top.Curr. Chem.* **2004**, *236*, 139-161.

[215] K. Nakatani, I. Saito, *Top.Curr. Chem.* **2004**, *236*, 163-186.

[216] M. Bixon, B. Giese, S. Wessely, T. Langenbacher, M. E. Michel-Beyerle, J. Jortner, *Proc. Natl. Acad. Sci. USA* **1999**, *96*, 11713-11716.

[217] H.-A. Wagenknecht, *Nat. Prod. Rep.* **2006**, *23*, 974-1006.

[218] J. Jortner, M. Bixon, T. Langenbacher, M. E. Michel-Beyerle, *Proc. Natl. Acad. Sci. USA* **1998**, *95*, 12759-12765.

[219] S. Delaney, J. K. Barton, *J. Org. Chem.* **2003**, *68*, 6475-6483;

[220] S. E. Rokita, T. Ito in *Charge Transfer in DNA: From Mechanism to Application* (Eds.: H. A. Wagenknecht), Wiley-VCH, **2005**, pp133-152.

[221] P. W. Anderson, *Phys. Rev.* **1959**, *115*, 2-13.

[222] J. Halpern, L. E. Orgel,*Discuss. Faraday Soc.* **1960**, 32-41.

[223] H. J. McConnell,*J. Chem. Phys.* **1961**, *35*, 508-515.

[224] F. D. Lewis, Y. Wu, *J. Photochem. Photobiol. C* **2001**, *2*, 1-16.

[225] E. Meggers, D. Kusch, M. Spichty, U. Wille, B. Giese, *Angew. Chem. Int. Ed. Engl.* **1998**, *37*, 460-462.

[226] C. A. M. Seidel, A. Schulz, M. H. M. Sauer, *J. Phys. Chem.* **1996**, *100*, 5541-5553.

[227] B. Giese, M. Spichty, *ChemPhysChem.* **2000**, *1*, 195-198.

[228] X. Li, Z. Cai, M. D. Sevilia, *J. Phys. Chem. A* **2002**, *106*, 9345-9351.

[229] J. Jortner, M. Bixon, A. A. Voityuk, N. Rösch, *J. Phys. Chem. A* **2002**, *106*, 7599-7606.

[230] M. Bixon, J. Jortner, *Chem. Phys. Special Issue on Molecular Wires* **2001**, *105*, 11057.

[231] M. Jakobsson, S. Stafström, *J. Chem. Phys.* **2008**, *129*, 125102.

[232] E. M. Boon, J. K. Barton, *Curr. Opin. Struct. Biol.* **2002**, *12*, 320-329.

[233] S. Anandan, M. Yoon, *Spectrochinmica Acta Part A*, **2004**, *60*, 885-888.

[234] R. Varghese, H-A.Wagenknecht, *Chem. Eur. J.* **2009**, *15*, 9307-9310.

[235] C. Holzhauser, S. Berndl, F. Menacher, M. Breunig, A. Göpferich, H.-A. Wagenknecht, *Eur. J. Org. Chem.* **2010**, 1239-1248.

[236] J. Mohanty, H. Pal, A. V. Sapre, *Photochem. Photobiol.* **2003**, *78*, 153-158.

[237] D. Rehm, A. Weller, *Ber. Bunsen-Ges. Phys. Chem.* **1969**, *73*, 834-839.

[238] S. Steenken, S. V. Jovanovic, *J. Am. Chem. Soc.* **1997**, *119*, 617-618.

[239] P.-G Wu, L. Brand, *Anal. Biochem.* **1994**, *218*, 1-13.

[240] J. R. Lakowicz in *Principles of Fluorescence Spectroscopy* 3rd, Springer Singapore, **2006**, pp. 445.

[241] A. P. Demchenko in *Advanced Fluorescence Reporters in Chemistry and Biology I* (Eds.: O. S. Wolfbeis), Springer Heidelberg, **2010**, pp. 14.

[242] V. Sartor, E. Boone, G. B. Schuster, *J. Phys. Chem. B* **2001**, *105*, 11057.

[243] A. A. Voityuk, N. Rösch, M. Bixon, J. Jortner, *J. Phys. Chem. B* **2000**, *104*, 9740-9745.

[244] M. A. O'Neill, J. K. Barton in *Topics in Current Chemistry: Long-Range Electron Transfer in DNA I*, (Eds.: G. B. Schuster) Springer New York, **2004**, pp. 105-106.

[245] a) Ethidium (donor) /7-deazaguanine (acceptor) in C. Wan, T. Fiebig, O. Kelley, C. R. Treadway, J. K. Barton, A. H. Zewail, *Proc. Natl. Acad. Sci. USA* **1999**, *96*, 6014-6019. b) 9-Amino-6-chloro-2-methoxyacridine (donor) /7-deazaguanine (acceptor) in S. Hess, M. Götz, W. B. Davis, M. E. Michel-Beyerle, *J. Am. Chem. Soc.* **2001**, *123*, 10046-10055. c) 2-aminopurine (donor) /inosine (acceptor) in M. A. O'Neill, J. K. Barton, *J. Am. Chem. Soc.* **2004**, *126*, 11471-11483.

[246] S. H. Choi, C. D. Frisbie in *Charge and Exciton Transport through Molecular Wires* (Eds.: L. D. A. Siebbeles and F. C. Grozema), Wiley-VCH, Weinheim, **2011**, pp. 63.

[247] S. L. Tesar, J. M. Leveritt III, A. A. Kumosov, A. L. Burin, *Chem. Phys.* **2012**, 393, 13-18.

[248] G. S. Blaustein, F. D. Lewis, A. L. Burin, *J. Phys. Chem. B* **2010**, *114*, 6732-6739.

[249] A. Cser, K. Nagy, L. Biczók, *Chem. Phys. Lett.* **2002**, *360*, 473-478.

[250] S. Anandan, M. Yoon, *Spectrochinmica Acta Part A* **2004**, *60*, 885-888.

[251] R. Ahlrichs, M. Bär, M. Häser, H. Horn, C. Kölmel, *Chem. Phys. Lett.* **1989**, *162*, 165.

[252] O. Treutler, R. Ahlrichs, *J. Chem. Phys.* **1995**, *102*, 346.

[253] K. Eichkorn, O. Treutler, H. Oehm, M. Haeser, R. Ahlrichs, *Chem. Phys. Lett.* **1995**, *240*, 283.

[254] K. Eichkorn, O. Treutler, H. Oehm, M. Haeser, R. Ahlrichs, *Chem. Phys. Lett.* **1995**, *242*, 652.

[255] K. Eichkorn, F. Weigend, O. Treutler, R. Ahlrichs, *Theo. Chem. Acc.* **1997**, *97*, 119.

[256] F. Weigend, *Phys. Chem. Chem. Phys.* **2006**, *8*, 1057.

[257] H. Horn, R. Ahlrichs, *J. Chem. Phys.* **1992**, *97*, 2571.

[258] R. A. Kendall, T. H. Dunning, Jr., R. J. J. Harrison, *Chem. Phys.* **1992**, *96*, 6796.

[259] F. Furche, D. Rappoport, ch. III of *"Computational Photchemistry"* in *Theoretical and Computational Chemistry Vol. 16* (Eds.: M. Olivucci), Elsevier, Amsterdam, **2005**.

[260] R. Bauernschmitt, R. Ahrichs, *Chem. Phys. Lett.* **1996**, *256*, 454.

[261] R. Bauernschmitt, R. Ahrichs, *J. Chem. Phys.* **1996**, *104*, 9047.

[262] A. Khan, *J. Chem. Phys.* **2008**, *128*, 075101.

[263] A. Khan, *Chem. Phys. Lett.* **2010**, *486*, 154-59.

[264] K. Kawai, T. Takada, S. Tojo, T. Majima, *J. Am. Chem. Soc.* **2003**, *125*, 6842.

[265] T. Takada, K. Kawai, X. Cai, A. Sugimoto, M. Fujitsuka, T. Majima, *J. Am. Chem. Soc.* **2004**, *126*, 1125-1129.

[266] A. K. Azad, R. P. Prasankumar, D. Talbayev, A. J. Taylor, R. D. Averitt, J. M. O. Zide, H. Lu, A. C. Gossard, J. F. O'Hara, *Appl. Phys. Lett.* **2008**, *93*, 121108.

[267] F. Wang, Y. Han, C. S. Lim, Y. H. Lu, J. Wang, J. Xu, H. Y. Chen, C. Zhang, M. H. Hong, X. G. Liu, *Nature*, **2010**, *463*, 1061.

[268] Y. J. Sun, Y. Chen, L. J. Tian, Y. Yu, X. G. Kong, Q. H. Zeng, Y. L. Zhang, H. Zhang, *J. Lumin.* **2008**, *128*, 15.

[269] T. Tachikawa, T. Ishigaki, J.-G. Li, M. Fujitsuka, T. Majima, *Angew. Chem. Int. Ed.* **2008**, *47*, 5348.

[270] B. Hillebrands, K. Ounadjela in *Spin Dynamics in Confined Magnetic Structures II*. Topics in Applied Physics 87, Springer, Berlin Heidelberg, **2003**.

[271] A. Baniodeh, I. J. Hewitt, V. Mereacre, Y.H. Lan, G. Novitchi, C. E. Anson, A. K. Powell, *Dalton Trans.* **2011**, *40*, 4080-4086.

[272] V. Mereacre, A. Baniodeh, C. E. Anson, A. K. Powell, *J. Am. Chem. Soc.* **2011**, *133*, 15335-15337.

[273] G. K. Liu, X. Y. Chen in *Handbook on the Physics and Chemistry of Rare Earths, Vol. 37: Optical Spectroscopy* (Ed.: K. A. Gschneidner, Jr., J.-C. G. Bünzli, V. K. Pecharsky), Elsevier Science Amsterdam, **2007**, pp. 99.

[274] T. K. Ronson, T. Lazarides, H. Adams, S. J. A. Pope, D. Sykes, S. Faulkner, S. J. Coles, M. B. Hursthouse, W. Clegg, R. W. Harrington, M. D. Ward, *Chem. Eur. J.* **2006**, *12*, 9299-9313.

[275] A. Baniodeh in *Cooperative Effects in Non-cyclic and Cyclic Fe/4f Coordination Clusters*, Dissertation, Karlsruhe Institute of Technology (KIT), **2012**.

[276] T. J. A. Wolf, D. Voll, C. Barner-Kowollik, A.-N. Unterreiner, *Macromolecules*, **2010**, *45*, 2257-2266.

[277] L. F. Jones, P. Jensen, B. Moubaraki, K. J. Berry, J. F. Boas, J. R. Pillbrow, K. S. Murray, *J. Mater. Chem.* **2006**, *16*, 2690-2697.

[278] S. K. Langley, B. Moubaraki, C. M. Forsyth, I. A. Gass, K. S. Murray, *Dalton Trans.* **2010**, *39*, 1705-1708.

[279] N. J. Cherepy, D. B. Liston, J. A. Lovejoy, H. Deng, J. Z. Zheng, *J. Phys. Chem. B* **1998**, *102*, 770-776.

[280] Y. P. He, Y. M. Miao, C. R. Li, S. Q. Wang, L. Cao, S. S. Xie, G. Z. Yang, B. S. Zou, C. Burda, *Phys. Rev. B* **2005**, *71*, 125411.

[281] S. Morup, E. Tronc, *Phys. Rev. Lett.* **1994**, *72*, 3278.

[282] D. M. Sherman, T. D. Waite, *Am. Mineral.* **1985**, *70*, 1262.

[283] M. P. Dare-Edwards, J. B. Goodenough, A. Hamnett, P. R. Trevellick, *J. Chem. Soc., Faraday Trans. 1* **1983**, *79*, 2027.

[284] Z. Zhang, C. Boxall, G. H. Kelsall, *Colloids Surf. A* **1993**, *73*, 145.

[285] L. Gomes, J. Lousteau, D. Milanese, G. C. Scarpignato, S. D. Jackson, *J. Appl. Phys.* **2012**, *111*, 063105.

[286] S. Tanabe, X. Feng, T. hanada, *Opt. Lett.* **2000**, *25*, 817-819.

[287] S. D. Jackson, T. A. King, *Opt. Lett.* **1998**, *23*, 1462-1464.

[288] J. L. Doualan, S. Girard, H. Haquin, J. L. Adam, J. Montagne, *Opt. Mater.* **2003**, *24*, 563-574.

[289] S. W. Magennis, A. J. Ferguson, T. Bryden, T. S. Jones, A. Beeby, I. D. W. Samuel, *Synthetic Metals*, **2003**, *138*, 463-469.

[290] M. Xiao, P. R. Selvin, *Rev. Sci. Instrum.* **1999**, *70*, 3877-3881.

[291] S. Lis, T. Kimura, Z. Yoshida, *J. Alloy. Compd.* **2001**, *323-324*, 125-127.

[292] T. Nishioka, J. Yuan, K. Matsumoto in *BioMEMS and Biomedical Nanotechnology Vol. III* (Ed.: M. Ozkan, M. J. Heller), **2007**, pp. 437-446.

[293] B. C. Faust, M. R. Hoffmann, D. W. Bahnemann, *J. Phys. Chem.* **1989**, *93*, 6371.

[294] K. Korobchevskaya, C. George, A. Diaspro, L. Manna, R. Cingolani, A. Comin, *Appl. Phys. Lett.* **2011**, *99*, 011907.

[295] A. G. Joly, J. R. Williams, S. A. Chambers, G. Xiong, W. P. Hess, *J. Appl. Phys.* **2006**, *99*, 053521.

[296] J. Kiwi, N. Denisov, Y. Gak, N. Ovanesyan, P. A. Buffat, E. Suvorova, F. Gostev, A. Titov, O. Sarkisov, P. Albers, V. Nadtochenko, *Langmuir* **2002**, *18*, 9054-9066.

[297] H. M. Fan, G. J. You, Y. Li, Z. Zheng, H. R. Tan, Z. X. Shen, S. H. Tang, Y. P. Feng, *J. Phys. Chem. C* **2009**, *113*, 9928-9935.

[298] R. W. Lof, M. A. van Veenendaal, B. Koopmans, H. T. Jonkman, G. A. Sawatzky, *Phys. Rev. Lett.* **1992**, *68*, 3924-3926.

[299] C. Klingshirn in *Semiconductor Optics*, 3rd, Springer, Berlin, **2007**.

[300] W. L. Kalb, S. Haas, C. Krellner, T. Mathis, B. Batlogg, *Phys. Rev. B* **2010**, *81*, 155315.

[301] A. Ellens, H. Andres, A. Meijerink, G. Blasse, *Phys. Rev. B* **1997**, *55*, 173-179.

[302] A. Ellens, H. Andres, M. L. H. ter Heerdt, R. T. Wegh, A. Meijerink, G. Blasse, *Phys. Rev. B* **1997**, *55*, 180-186.

[303] K. A. Gschneider, L. Eyring in *handbook on the Physics and Chemistry of rare earths Vol. 1: metals*, North-Holland Publishing Company, **1978**.

[304] S. D. Barrett in *The Structure of Rare-earth Metal Surfaces*, Imperial College Press, **2001**.

[305] R. S. Meltzer, W. M. Yen, H. Zheng, S. P. Feofilov, M. J. Dejneka, B. M. Tissue, H. B. Yuan, *Phys. Rev. B* **2001**, *64*, 100201.

[306] J. Tauc in *Amorphous and Liquid Semiconductors*, Plenum, New York, **1974**, pp. 159.

[307] S. K. Mehta, S. Chaudhary, S. Kumar, S. Singh, *J. Nanopart. Res.* **2010**, *12*, 1697.

[308] A. A. Tahir, K. G. U. Wijayantha, S. Saremi-Yarahmadi, M. Mazhar, V. McKee, *Chem. Mater.* **2009**, *21*, 3763-3772.

[309] J. C. Phillips in *Bonds and Bands in Semiconductors*, Academic Press, New York and London, **1973**, pp. 3.

[310] S. Baskoutas, A. F. Terzis, *J. Appl. Phys.* **2006**, *99*, 013708.

[311] T. Ihn in *Semiconductor Nanostructures*, Oxford University Press, New York, **2010**.

[312] P. Y. Yu, M. Cardona in *Fundamentals of Semiconductors – Physics and Materials Properties* 4th, Springer Berlin Heidelberg, **2010**.

[313] S. Kan, T. Mokari, E. Rothenberg, U. Banin, *Nature*, **2003**, *2*, 155.

[314] G. Ouyang, X. L. Li, X. Tan, G. W. Yang, *Appl. Phys. Lett.* **2006**, *89*, 031904.

[315] C. Q. Sun, *Prog. Solid State Chem.* **2007**, *35*, 1.

[316] N. M. Dimitrijević, D. Savić, O. I. Mićić, A. J. Nozik, *J. Phys. Chem.* **1984**, *88*, 4278-4283.

[317] R. Bosînceanu, N. Suliţanu, *J. Optoelectron. Adv. Mater.* **2008**, *10*, 3482-3486.

[318] G. V. Manohara, S. V. Prasanna, P. Vishnu Kamath, *Eur. J. Inorg. Chem.* **2011**, 2624-2630.

[319] A. L. Crumbliss, P. S. Lugg, N. Morosoff, *Inorg. Chem.* **1984**, *23*, 4701.

[320] T. Uemura, M. Ohba, S. Kitagawa, *Inorg. Chem.* **2004**, *43*, 7339.

[321] D. C. Arnett, P. Vöhringer, N. F. Scherer, *J. Am. Chem. Soc.* **1995**, *117*, 12262-12272.

[322] D. F. Watson, A. B. Bocarsly in *Femtochemistry and Femtobiology: Ultrafast Dynamics in Molecular Science* (Ed.: A. Douhal, J. Santamaria), World Scientific Publishing, **2002**, pp. 664.

[323] B. P. Macpherson, P. V. Bernhardt, A. Hauser, S. Pagès, E. Vauthey, *Inorg. Chem.* **2005**, *44*, 5530-5536.

[324] M. Thoss, H. Wang, *Chem. Phys. Lett.* **2002**, *358*, 298-306.

[325] D. Weidinger, D. J. Brown, J. C. Owrutsky, *J. Chem. Phys.* **2011**, *134*, 12450.

[326] H. Kamioka, Y. Moritomo, W. Kosaka, S Ohkoshi, *Phys. Rev. B* **2008**, *77*, 180301.

Acknowledgements

I would like to acknowledge a lot of nice people for the helps in this work. First of all, I want to thank my advisor PD Dr. A.-N. Unterreiner, who was always available and helpful to show me different ways to approach a research goal. Without his support I won't be able to pursue these studies.

My special thanks go to Dr. O. Schalk, from whom I gave lots of intensive supports, constructive discussions and suggestions.

During my work I closely worked with my colleagues, Dr. M. Klinger, Dr. T. Wolf and H. Ernst in our group. They gave me experience, advices, new skills and overall friendship and good working atmosphere. I want to thank all the members of the molecular physical chemistry group with whom I spent a great time in Karlsruhe.

I'm also grateful to M. Bradler, Dr. A. Baniodeh, Prof. Dr. T. S. Balaban, Prof. Dr. H.-A. Wagenknecht, Prof. Dr. A. K. Powell and Prof. Dr. E. Riedle for the cooperation and their scientific advices.

I would like to say my words of appreciation to D. Kelly and P. Hibomvschi for their help in technical and administrative issues.

My deepest thanks are for my family, Ying, Benben, my parents, and other who helped me and cared about me through all these years, without their support I won't finish my PhD.

List of Publications

Publications:

- O. Schalk, **Y. Liang**, A.-N. Unterreiner, On Ligand Binding Energies in Porphyrinic Systems, *Z. Phys. Chem.* **2013**, *227*, 35-48.

- T. Ehrenschwender, **Y. Liang**, A.-N. Unterreiner, H.-A. Wagenknecht, T. J. A. Wolf, Fluorescence Quenching over Short Range in a Donor-DNA-Acceptor System, *ChemPhysChem.* **2013**, *14*, 1197-1204.

- **Y. Liang**, O. Schalk, A.-N. Unterreiner, Transient Anisotropy in Degenerate Systems: Experimental Observation in a Cd-porphyrin, *EPJ Web of Conferences*, **2013**, *41*, 05014.

- **Y. Liang**, M. Bradler, M. Klinger, O. Schalk, M. C. Balaban, T. S. Balaban, E. Riedle, A.-N. Unterreiner, Ultrafast Dynamics of Meso-tetraphenylmetalloporphyrins: The Role of Dark States, *ChemPlusChem.* **2013**, to be submitted.

Conference Contributions:

- Y. Liang, O. Schalk, A.-N. Unterreiner, Ultrafast Relaxation Dynamics of 5,10,15,20-Tetraarylporphyrin and Metallanaloga, Bunsentagung 2011, 110th Annual German Conference on Physical Chemistry, 2011, Poster.

- Y. Liang, O. Schalk, A.-N. Unterreiner, Transient Anisotropy in Degenerative Systems: Experimental Observation in a Cadmium-Porphyrin, XVIII International Conference on Ultrafast Phenomena, 2012, Poster.